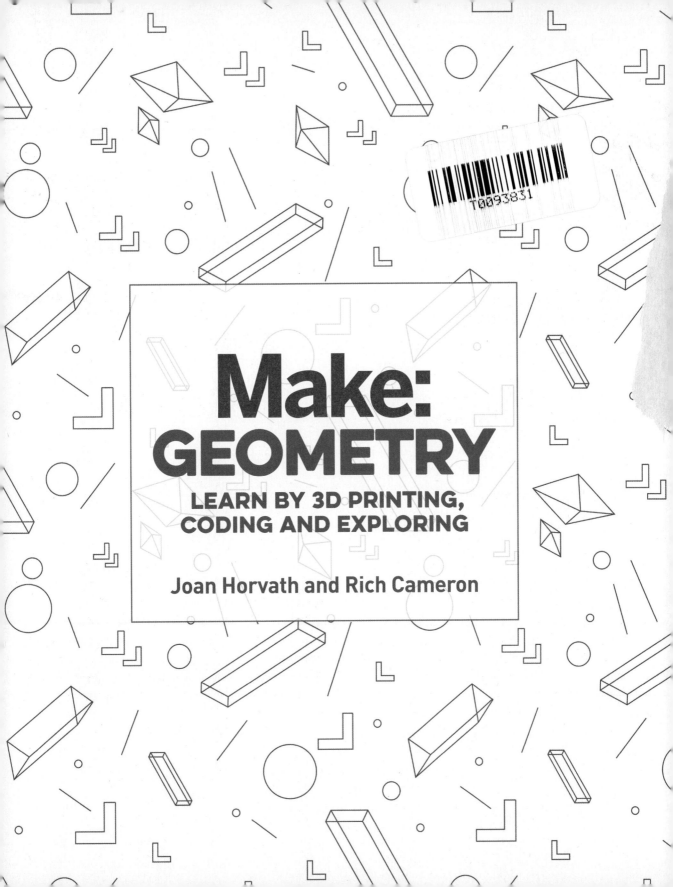

Make:
GEOMETRY

LEARN BY 3D PRINTING, CODING AND EXPLORING

Joan Horvath and Rich Cameron

Make: GEOMETRY

By Joan Horvath and Rich Cameron

Published by Make: Community LLC

150 Todd Road, Suite 200, Santa Rosa, CA 95407

Make: books may be purchased for educational, business, or sales promotional use.
Online editions are also available for most titles.
For more information, contact our corporate/institutional
Sales department: 800-998-9938

Publisher: Dale Dougherty
Editor: Patrick Di Justo
Creative Director: Juliann Brown
Design: Jason Babler

This book's body font is Din Pro Light. Subheads are set in Geogrotesque Condensed.

First Edition, July 2021

See www.oreilly.com/catalog/errata.csp?isbn=9781680456714 for release details.

ISBN: 9781680456714

(2022-01-07)

O'REILLY ONLINE LEARNING

For more than 40 years, www.oreilly.com has provided technology and business training, knowledge, and insight to help companies succeed.

Our unique network of experts and innovators share their knowledge and expertise through books, articles, conferences, and our online learning platform. O'Reilly's online learning platform gives you on-demand access to live training courses, in-depth learning paths, interactive coding environments, and a vast collection of text and video from O'Reilly and 200+ other publishers. For more information, please visit www.oreilly.com

HOW TO CONTACT US:

Please address comments and questions concerning this book to the publisher:

Make: Community LLC

150 Todd Road, Suite 200, Santa Rosa, CA 95407

You can also send comments and questions to us by email at books@make.co.

Make: Community is a growing, global association of makers who are shaping the future of education and democratizing innovation. Through *Make:* magazine, and 200+ annual Maker Faires, *Make:* books, and more, we share the know-how of makers and promote the practice of making in schools, libraries and homes.

To learn more about *Make:* visit us at make.co.

TABLE OF CONTENTS

CHAPTER 4: CONSTRUCTIONS 62

CHAPTER 5: THE TRIANGLE BESTIARY 74

PREFACE

Geometry, of all the branches of mathematics, is the one that is most easily visualized by making something. However, it is all too easy to reduce it to reams of formulas to memorize and proofs to replicate. A lot of the basics of geometry are a few thousand years old, and mostly solve practical problems. We will try to get at both the practicality of geometry, without losing the puzzle-solving and aesthetics that also make it joyful to learn.

WHO THIS BOOK IS FOR

This book is for anyone learning geometry, or perhaps for those of you who learned geometry a long time ago and are trying to repress painful memories. This book is intended for parents and teachers of children learning middle- and high-school geometry, in particular those who prefer to learn by making something. Many maker education projects are tacked on to existing curriculum and depend on the traditional way of teaching to impart core understanding.

This book, in contrast, starts with something the student can make or observe, and we deduce the principle in question from that. Students who might be good at math, but restless and bored when taught with lectures, will particularly benefit. We will also stretch beyond standard K-12 geometry into a few more sophisticated concepts that make things more interesting.

WHAT YOU'LL NEED

This book assumes a basic knowledge of algebra, at the level of understanding positive and negative numbers, fractions, and being able to solve for x in an equation like:

$3x + 5 = 4$

We explain concepts like raising a number to a power and taking a root, but of course, it is helpful if you have seen those before.

The models were written in the free, open-source CAD program OpenSCAD (OpenSCAD.org). If you don't have a 3D printer readily available, visualizing and manipulating the models in the OpenSCAD environment will provide some of the experience. OpenSCAD runs on laptop or desktop computers running Windows, macOS, or Linux. As of this writing, OpenSCAD does not run on Chromebooks or tablets. Creating models in OpenSCAD requires learning computer programming skills, but we give enough pointers that someone without that background should be able to manage the basics.

In some cases, though, we have several OpenSCAD models that are used together, and OpenSCAD doesn't have a good way to support that on the screen without printing out physical models. In those cases, you still should be able to read this book and understand the concepts from our photographs of our 3D printable models if using a 3D printer isn't an option. We encourage you, though, to print the models on any consumer-level filament-using 3D printer. As we discuss in Chapter 2, decent printers can be found in the under $500 range, and many community libraries have printers available.

Where possible, we also provide examples you can create with common household items, such as paper, a flashlight, rulers, measuring cups, or books. A protractor, a compass for drawing circles, and a compass of the north-finding variety (or compass app on a phone) will also come in handy as you go through the book. Other than a 3D printer, we have tried to stick to supplies that can be purchased cheaply or cobbled together.

TEACHING AND LEARNING WITH THIS BOOK

This book aims to take geometry back to its physical roots with 3D printed models, as well as demonstrations with household objects. Often, we'll mention the circumstances under which the math was invented. We find doing that helps us think about how to frame it for someone else who is seeing a concept for the first time.

Chapters 2 and 3 dive into OpenSCAD, a code-based computer-aided design (CAD) program. The models in the rest of the book are written in OpenSCAD. Since OpenSCAD requires that models be built in a programming language that is somewhat like C or C++, this book could also motivate a learn-to-code curriculum.

This is not a traditional geometry textbook. The book is designed to be read through starting at the beginning, but we have tried to cross-reference where we use material from other chapters if you want to dip around instead. We would suggest, though, reading Chapters 2 and 3 before continuing so you understand OpenSCAD and other background information. Chapters 3 through 10 build up an understanding of geometry concepts usually encountered in middle and high school, in some cases pointing out the road to concepts normally learned much later. The last two chapters of the book are projects that are a little more sophisticated, and which might be the basis for a science fair project or at least further exploration.

In addition to the material in this book, lesson plans designed for teachers of visually impaired and blind students that address some of the same models and topics are freely available at www.nonscriptum. com/geometry.

ALIGNMENT TO STANDARDS

Mathematics teaching standards vary by jurisdiction. Many of them in the US broadly follow the Common Core Standards. The Common Core geometry standards can be found at **http://www.corestandards.org/Math/Content/G/**. In the Appendix, we include a table of keywords and concepts that each chapter addresses. We also were influenced by the California math standards, available at **https://www.cde.ca.gov/be/st/Ss/documents/ccssmathstandardaug2013.pdf**.

Most of these standards start with abstract 2D reasoning and go on to 3D afterward, or focus on exhaustive analyses of, for example, all the ways one triangle can be congruent to another.

It is our belief that teaching with 3D objects first and then returning to abstractions (if at all) gives a better foundation and better intuition for higher-level math, like calculus and engineering applications. Thus we do not claim to cover all of any grade level's required math in any jurisdiction, but rather that we will cover key concepts that lend themselves to this approach.

Our approach follows ideas that have been called, among other things, constructivism, problem-based learning, and active learning. We have been influenced by Paul Lockhart's work, particularly his books, *The Mathematician's Lament* (Belknap:2009) and *Measurement* (Belknap:2012).

Of course, we are also not the first to create 3D printable math models. Many existing models are inhabitants of what we call "math zoos"—models that are of obscure but pretty functions, or (cringe) extruded versions of 2D projections of 3D objects. We present scaffolded learning that might not be in the "normal" sequence, but is suited to hands-on learning.

ACKNOWLEDGMENTS

The authors would like to thank the many people who commented on our early models and explanations. We owe a particular debt of gratitude to our colleagues and students at Pasadena's Institute for Educational Advancement, where we tried out some of our ideas in their earliest form on their ready-for-anything gifted students. We also appreciated the Playline community, run by AnnMarie Thomas' Playful Learning Lab at the University of St. Thomas, for being a sounding board and forum.

We also want to note that developing some of the material for this book was supported in part by grant number 90RE5024, from the U.S. Administration for Community Living, Department of Health and Human Services, Washington, D.C. 20201. We also are grateful for the insights of Yue-Ting Siu into creating tactile models and the Smith-Kettlewell Eye Institute for their support.

Astronomer Steve Unwin was very helpful in our framing of some of the astronomy concepts, as well as providing some photos in Chapter 13. He also happens to be Joan's husband and kept things going with pizza and sympathy at crucial stages of the project.

Last, but definitely not least, we thank all the staff at MAKE Community LLC who supported all the pieces that go into producing a book like this one. We particularly appreciated our editor, Patrick Di Justo, who provided our prose with a lot of tough love and said, "Not in this town," when we missed a definition or Oxford comma.

ABOUT THE AUTHORS

Joan Horvath and Rich Cameron have pioneered using 3D printing to teach math and science. In 2015 they formed Nonscriptum LLC (**www.nonscriptum.com**), a consulting and training partnership through which they help people use maker technologies to solve real-world problems.

Previous to this book, they have written seven books for the Apress imprint of Springer-Nature, including two books of 3D printable science projects. They are also the authors of many courses for LinkedIn Learning (formerly Lynda.com) covering a range of topics in additive manufacturing. In 2020, LinkedIn Learning launched their regular series, "Additive Manufacturing: Tips, Tricks and Techniques." Their books, courses, and projects are linked at www.nonscriptum.com.

Joan and Rich met at Deezmaker 3D Printers where they were part of its Kickstarter-era team. They have been around the Maker community since its very early days. Joan is a recovering rocket scientist who spent 16 years at JPL, followed by more than 20 more years as part of various entrepreneurial ventures. She is an alumna of MIT and UCLA. Rich is an experienced maker who developed the RepRap Wallace and Deezmaker Bukito 3D printers.

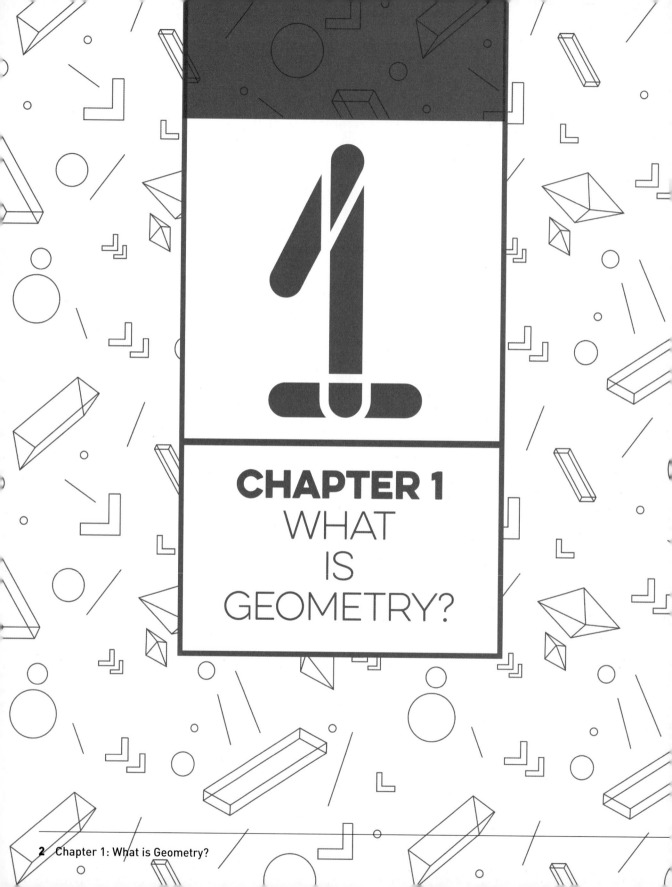

1

CHAPTER 1
WHAT
IS
GEOMETRY?

The word "geometry" has its roots in ancient Greek words *geos* for "earth" and *metria* for "measure". Over two thousand years later, we're still using concepts invented by Greek mathematicians and philosophers like Archimedes and Euclid. The discipline concerns itself with ways to construct and analyze the properties of different shapes, from one-dimensional points to three-dimensional objects. Put that way, geometry sounds pretty removed from daily life, but its basics underlie everything from how much paint to buy to cover a wall to predicting where the sun will be at local noon tomorrow.

Fundamentally, we think of geometry as taking apart seemingly simple things like circles and triangles, and seeing how to put them back together in creative ways. One of the challenges of getting under the hood on some of these seemingly simple concepts is that it can be difficult to define these common-place shapes, like a square or a circle. We'll look at ways to do that literally, with 3D prints, paper or string, and also with a dash of paper and pencil analysis. (Don't worry, we'll do that part with you.)

EUCLID

About 2400 years ago, the Greek mathematician Euclid wrote a book called "The Elements," which boiled geometry down to a few concepts and derived many others from those. At first hand-copied, then printed, it has formed the foundation for nearly two and a half millillenia's worth of geometry books (including this one). You can still buy (or download) a translated copy with or without various people's annotations.

For the most part, the geometry we will learn in this book is called "Euclidean geometry." Euclidean geometry can be divided into plane (2-dimensional) and solid (3-dimensional) concepts. We will jump back and forth to a degree, since manipulatives and 3D printer technology sometimes make it easier to see a 3D model first, and pick up the 2D ideas later.

BEYOND EUCLID

Europeans in the 1600s started expanding various aspects of Euclid's system, as we will see when we learn about Descartes and coordinate systems in Chapter 2. Descartes' coordinates let us label points on a shape, and not long after that, Newton and Leibnitz invented calculus which depended on that insight. (Newton also famously laid out his laws of physics and the calculus he invented for it in mostly geometrical terms.) Collectively, those ideas launched a field now called analytic geometry. This ties geometrical shapes to a coordinate system to let you compute things like the apparent path of the sun in the sky over the course of a year (as we see in Chapter 12).

In the early 1900s, people extended Euclid's way of thinking even more when they needed to start talking about more dimensions. Both Newton's physics and Euclid's geometry needed to be extended to let Einstein create his theory of relativity, which added the fourth dimension of time to the three Euclidian ones in space.

More recently, fields like computational geometry have arisen that are possible because computers are able to do fantastical amounts of number-crunching. Computer animation (originally based on simple geometric shapes) has also driven interest in many crossover fields between computer science and geometry to create imaginary worlds for virtual cameras.

THINK LIKE A MATHEMATICIAN

People often are afraid of math and think they can't learn it. Some people like to say it's like learning a language, but we think that's misleading and focuses too much on the words for things. Instead, we think that geometry is a way of creating ideal, simple versions of real-world objects. Once you have those, you can approximate what is going on in the real world by building from there.

When you draw a little picture for yourself to figure things out, or explain something to someone else with a sketch on a napkin, typically you just draw a few abstract squiggles. It's not a perfectly shaded artist's sketch (unless you're awfully good at that sort of thing). The real power of creating ideal circles, triangles, and so on (as Euclid realized) is that it's possible to show relationships among them and use them to think about more complicated abstract ideas. That in turn lets you see ways of solving practical problems that might have been hidden by all the details of the real world otherwise.

For example, in Chapter 7, you will learn to compute where you are on earth (your latitude and longitude) by measuring the length of the shadow of a stick at noon. We do this by overlaying the real world with an imaginary grid of lines and using where the sun is on the grid to deduce where we are. It took quite a few leaps of imagination 2400 years ago, along with some enhancements about 500 years ago and the development of good clocks, for us to develop that power. The ideas in this book are tools, just like a screwdriver, but easier to carry around. We'll try to show you when to use each one.

We also think that the best way to get some intuition about how to overlay these imaginary perfect objects on real ones is to make those objects and play with them. We will explore both digital and physical ways of doing that. We hope you'll try them all and see which one works best for you!

STRUCTURE OF THIS BOOK

The rest of the chapters in this book first introduce the 3D printable models (if any) used in that chapter, plus any other materials you might need, like a protractor or paper. Then, we develop a little bit about the history of the concept we are learning, and then get into using the physical models, or describe the project to do, as the case may be. Each chapter ends with a summary and suggestions on where to learn more, as well as answers to some questions posed during the chapter.

We have a mix of digital projects (requiring only OpenSCAD) and physical ones. Chapter 2 introduces the mechanics of using OpenSCAD and its workflow, and a brief introduction to 3D printing. In Chapter 3, we take a deeper dive into how to use OpenSCAD to simulate geometric transformations like rotation and translation, learn about general 2D shapes (polygons), and are introduced to 3D ones (polyhedrons). We wind up those two chapters with a project to create a 3D printable castle to try out these concepts. You can choose to keep this an all-digital exercise and just see your castle on the screen, or actually print your creation if you have access to a 3D printer.

Then, for a change of pace, in Chapter 4 we learn how to use a drawing compass to do traditional constructions. We use just paper, pencil, a straightedge, and a compass to create various shapes and see the relationships among them. The latter part of Chapter 4 uses one of these traditional constructions as a starting point for an OpenSCAD model to tie those pieces together.

Chapter 5 is all about triangles, with an assortment of 3D printable models (some of which can be created with paper instead). For this and the rest of the chapters in the book, we show concepts with a mix of 3D prints and lower-tech demonstrations where possible.

Next Chapter 6 builds on Chapter 5's introduction to triangles to learn the Pythagorean Theorem, which lets us figure out some properties of triangles with one 90-degree angle. We get introduced to some basic trigonometry, too, which we'll need later in the book.

We tie a lot of these pieces together in Chapter 7, where we learn about circles and the relationships among triangles, polygons, and circles. We apply what we learn to figure out where we are on earth (our latitude and longitude) based on just the length of a particular shadow measured at noon.

We learn a lot about how to calculate 2D objects in the early chapters. Chapter 8 takes us into the third dimension to learn about volume. We look into some practical applications of combining volume with the physics concepts of density (how much a given volume of a material weighs) and displacement (how much volume of water a given floating mass will move out of the way). We'll use all that to recreate a version of a legendary experiment to use these relationships to deduce what material something is made of.

In Chapter 9, we create 2D shapes that will exactly cover 3D objects, known as nets. These let us find out how much area the surface of a 3D shape would have if we could flatten it out into a 2D one. Chapter 10 shows us how to 3D print models of a prism, pyramid, cylinder, sphere, or cone that have been sliced at various angles. Many of these slices have surprising properties, particularly the slices through a cone (called "conic sections") which we introduce in the chapter.

Finally, in Chapters 11, 12, and 13, we walk you through some larger projects that expand upon the ideas in the earlier chapters. Chapter 11 details unconventional ways to construct ellipses, parabolas, and hyperbolas, and draws physical insights from those constructions. Chapter 12 extends geography and astronomy concepts from Chapter 7, and will let you explore how the path of the sun varies over a year and learn about the geometry of sundials. We call Chapter 13 the Geometry Museum, and review more open-ended projects in the architecture or aesthetic realm, ending with a few geometrical phenomena that you can explore more on your own.

SUMMARY AND LEARNING MORE

In this chapter, we gave you a brief introduction to what geometry is and how it fits into the bigger picture of mathematics. We introduced Euclid (and that this book is about Euclidean geometry for the most part).

Then we surveyed the book so that you can decide where you want to start. This book is not intended to be a full textbook for either middle or high school geometry. Rather, we imagine that you are learning these concepts from a more traditional textbook, and the material here can be used to enhance your understanding. Or, perhaps, if you find the material hard to learn the traditional way, perhaps learning by making with the models and projects in this book will suit you better.

We'll give you pointers to other resources as we go forward. If you want more about any particular geometry topic, you can try the wonderful videos at the Khan Academy (www.khanacademy.org). Less systematically, there are also quite a few mathematicians with channels on YouTube or other platforms. We are very partial to the YouTube channels run by creators Numberphile and 3Blue1Brown, both of whom have a very intuitive approach to mathematics.

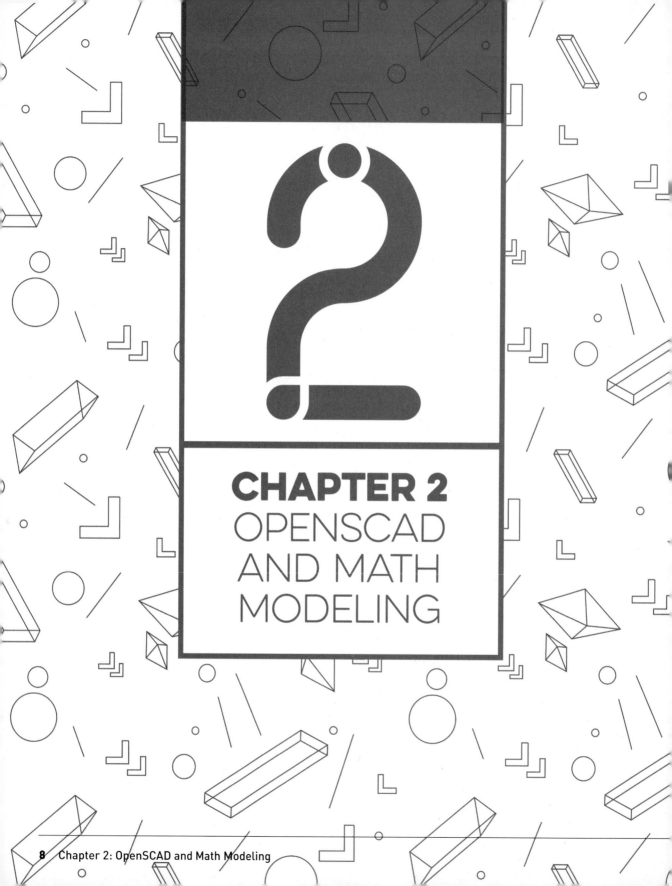

CHAPTER 2
OPENSCAD
AND MATH
MODELING

This chapter introduces the basics of using OpenSCAD to model math and science concepts, and ties it in with how to think about (and encode) core ideas in geometry. OpenSCAD is a computer-aided design (CAD) program. There are many different types of CAD. OpenSCAD is a *constructive solid geometry* program. That means that OpenSCAD lets you define basic 2D and 3D geometric shapes (like spheres, cylinders, and so on) and then create more complex objects by adding, subtracting, and intersecting the basic shapes.

We use OpenSCAD for the 3D printable models in this book. Beyond that, though, OpenSCAD is also a good visualization tool. Many of the basic manipulations in geometry are simple commands in Open-SCAD, and in Chapter 3 we will encourage you to do some experiments in OpenSCAD to develop your intuition about the basic manipulations of geometry. OpenSCAD is a free, open-source program, which you can download at **www.openscad.org**. There is a very good manual available on the site as well. As of this writing, it runs on Mac, Windows, and Linux computers, but not on tablets or Chromebooks. Marius Kintel is the primary developer and maintainer of this open-source project.

OpenSCAD models are written in a computer language that is similar to the C, Java, or Python languages. There is no drag and drop capability; everything is written as text. There are several visual-programming subsets of OpenSCAD out there, but these variants may not have the full functionality used in our models.

In this chapter, we introduce OpenSCAD and explain where you can get the models that we use throughout the book. You don't need to learn to write your own models from scratch to use this book, but you will learn more if you get comfortable enough to alter ours.

We also introduce Tinkercad, a free, introductory CAD program from Autodesk that allows you to drag and drop shapes. It is challenging to be very precise with Tinkercad, but you can have some fun exploring the basics with it.

THE MODELS

The models in this book are designed to be 3D printed. If you don't have access to a 3D printer, we've suggested alternatives in some cases (for example, making a rough equivalent in 2D with paper). In a lot of cases, you can get at least some of the intuition by opening the models in OpenSCAD and rotating them around to get a good look. Some models are meant to fit together or to be puzzles, and those are really dependent on being created physically.

We have tried hard to make the models so that they will print on a printer that is perhaps not perfectly tuned. Prints should work with basic settings on most machines. For the few prints that are a bit more aggressive, we have given some suggestions to improve the probability of success.

SOME MODELS HAVE SMALL PARTS

These are educational models, and some have small parts. They are intended for middle and high

FIGURE 2-1: A 3D printer, annotated.

school students, and should not be used around very young children. Some models won't print correctly if they are scaled up or down. Trying to make parts smaller might make them unprintable, while increasing the size may make things fit together too loosely. As we go along, we note which models cannot be scaled from a printability point of view.

3D PRINTING

If you've never tried 3D printing, you might wonder how it all works. The models in this book are designed to print on a 3D printer that uses *filament* as its working material. Filament is a plastic string, typically sold in 1kg spools. A 3D printer melts the filament and extrudes it through a fine nozzle, typically between 0.3 and 0.8mm in diameter. A 3D object is built up one layer at a time on a build platform from this extruded plastic (Figure 2-1).

There are many different designs of 3D printer. Some keep their build platform stationary, and have the extruder move in all three dimensions. In others, the build platform moves in one dimension and the extruder head moves in two, and so on. In the case of the 3D printer in Figure 2-1 (designed by Rich, as it happens) the platform moves in one horizontal dimension, and the extruder is on a gantry that moves vertically and also in the other horizontal dimension.

Many people describe a 3D printer as a robotic hot glue gun, which is really a better analogy than is a paper printer. There is no particular relationship between printers on paper and 3D printers.

There is also quite a bit of software involved. Here is how the software side of the 3D printing process works (Figure 2-2)

FIGURE 2-2: The 3D printing software flow

First, a user creates a 3D CAD model (for example, in OpenSCAD). Models can be saved in several formats in OpenSCAD. Models saved as .scad files can be edited in OpenSCAD, but are not ready to 3D print. For 3D printing, you have to "render" a model, which changes the model from a set of geometric shapes that have been combined in various ways into a surface covered with triangles. This file is called an .stl file, and the process of covering a surface with triangles is called *tessellation*.

STL stands for either STereoLithography or Standard Tessellation Language, depending on who you ask. Stereolithography was the first deployed 3D printing technology, about 40 years ago, by Chuck Hull of 3D Systems. (3D printers that create objects using lasers to harden liquid resin one layer at a time still use stereolithography.) A 3D model is represented in an .stl file as a long, long list of triangles and information on how those triangles are oriented. See the section "3D Printing Meshes" in Chapter 5 for more about these triangles. The last point in the 3D printing software chain that is printer-independent is the .stl file.

Once you have an .stl file, you will run a *slicing* computer program. Sometimes also called a "slicer," this program takes in an .stl file and produces a file of commands for a printer to execute, one layer at a time. The slicer has a variety of settings, like the temperature of the nozzle, how thick each layer is, and so on. The slicer's output will be just for one kind of 3D printer, one particular material, and one CAD model.

For most 3D printers, the instructions that come from the slicer will be in a format called *G-code*, though some printers use proprietary alternatives. Putting G-code into the printer often involves putting a G-code file on an SD card or USB drive and plugging it into the printer, then selecting the file from the printer's on-screen menu. Some printers also use a Wifi connection to exchange files.

Since this is not primarily a book about 3D printing, we are not going to go more deeply into the process of printing. If you need more background, we've written several books on 3D printing. You might particularly find our 2020 book, *Mastering 3D Printing, 2nd Edition* (NY: Apress) or Liza and Nick Kloski's *Getting Started with 3D Printing, 2nd Edition* (2021, Make Community LLC) good resources if you are new to 3D printing or want to up your game.

You may be lucky enough to have a printer at home. Many public libraries have printers available for patron use, typically after some basic training and perhaps for a small fee. If you are a student, your school might have some printers, too. MAKE Magazine routinely reviews printers if you want some advice on buying one. There are decent ones now for under $500; read reviews and know your limitations if you are thinking about saving money with a kit. We have designed the models so they will fit on the smaller end of consumer printers (ones about 150mm in each dimension).

MATERIALS

All of the models in the book will print fine in basic PLA (polylactic acid), one of the cheapest and most common 3D printing materials, which is typically made from corn or sugar cane. You can typically buy basic PLA for about $15 a kilogram, which would probably be enough to print everything in the book with some left over. For aesthetics, we used translucent or transparent PLA for some prints, which tends to be a little pricier, or transparent PETG (polyethylene terephthalate glycol, a modified formulation of the plastic commonly used for soda bottles) which can be a little harder to work with. Models with a shiny surface were printed with "silk PLA," which is PLA with

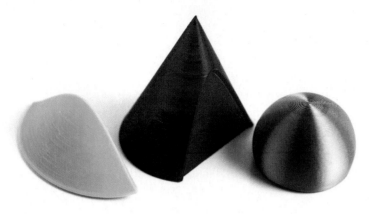

FIGURE 2-3: Three prints (L to R): in basic PLA, PETG, and silk PLA

some lignin fibers that tend to hide the layer lines and give a smoother-looking finish (Figure 2-3).

DOWNLOADING THE MODELS: GITHUB

The repository **https://github.com/whosawhatsis/Geometry** has all the models for this book. Github is a site that has many repositories, which are collections of freely available computer models, software, and so on. This repository includes the models for this book as well as a few that are used in other, overlapping projects of ours. We also plan to keep adding to it, so there may be some bonus models in there which are not described in the book. The start of each chapter has a sidebar that lists which models (and other supplies) are used in the chapter.

To download all the models, go to the link we just noted, and click the "Release" button. That will download the latest version of all editable OpenSCAD files, plus select STL files. In many cases, the text will suggest you play around with a lot of variations, and as a practical matter, we give you just the STL files that make representative cases, usually the prints used in the Figures. You will need to make STL files for the rest in OpenSCAD, which we will describe shortly. Alternatively, you can click on just one model in the list on the repository page. It will open the OpenSCAD text file, which you can select, copy and paste into OpenSCAD. Note that the Release includes models described in this book and more, since the repository also supports other projects (see the Appendix for more).

The models are released under a Creative Commons Attribution 4.0 International license. (You can read more about these licenses on creativecommons.org.) This license means that you can do whatever you

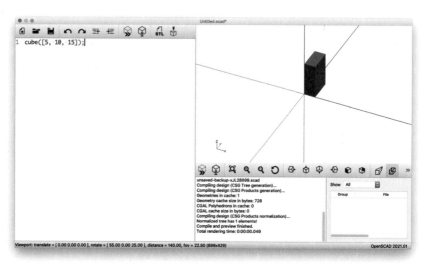

FIGURE2-4: OpenSCAD basic interface.

want with the models—print them, be paid by someone else to print them, create variations—as long as you attribute them to us. Each model has a few comment lines at the top that give the language we ask that you use for attribution.

In this chapter, we will be showing you snippets of a few lines to demonstrate various features. These are either in screenshots or in the text. We have not put these in the repository, but you can easily type them into OpenSCAD.

Note that Github requires that users be at least 13 years old, so if you are under 13, please ask an adult to download the files for you.

OPENSCAD

To use OpenSCAD, first download it from openscad.org and follow any on-screen instructions to install it. The manual (under the "documentation" tab on the website) has a very good and inclusive description of how to use the program and its associated modeling language. We will summarize the key points here, but we suggest you refer to the manual as well. We will show menu items and OpenSCAD commands in **a different font**, here and elsewhere in the book.

The program has three basic panes (Figure 2-4). The editor is where you type (or paste in) the text of your model. The *console* will tell you if you have issues when your print is rendered and is also where any messages that you write into your models will appear. Finally, there is a *display window* that displays the model when you preview or render it. Everything in OpenSCAD has to be input as text; there is no edit capability in the display window (but you can rotate the view and zoom around in it). You can rearrange the windows relative to each other as you like by dragging on their top bars.

OPENSCAD WORKFLOW

We will describe how to use the basic functionality with the menus arrayed across the top of the OpenSCAD screen. There are also various buttons and shortcuts described in the manual. We use the convention **Menu > Item** to say "Click on **Menu**, then click on the **Item** shown in the pulldown menu that appears."

- Since OpenSCAD does crash sometimes, it is prudent to use **File > Save** before trying any of the other functions here to save an OpenSCAD editable (.scad) file. OpenSCAD does not autosave, and it is easy to lose your entire model if the program crashes or hangs during preview or render.
- To preview (display a fast, imperfect version of a model), use **Design > Preview**.
- To render (creates an accurate model, but usually takes longer than the preview; required for exporting an STL) a model, use **Design > Render**.
- To export an STL file after rendering, use **File > Export > Export as STL**.
- If you can't see one or more of the panes, go to the **View** menu and make certain that the respective windows are checked.
- To display and then export an STL file by copying and pasting from Github, use these steps:
  - **File > New,** then paste in the text from the Github model
  - **File > Save As...**
  - **Design > Preview**
  - Assuming you like what you see, **Design > Render**
  - **File > Export > Export as STL**

Finally, OpenSCAD has example models (not all of which are printable) that you can find in **File > Examples**.

NAVIGATING ON THE SCREEN

Once you have previewed or rendered a model, you can "fly around" it to view it in 3D. Click on the display window, and then use your mouse or trackpad as follows.

- To move around in general, hold down your left mouse button

(Windows) or single mouse button (Mac) and move around as you desire.
- To shift the field of view, right-click (Windows) or Control-click (Mac).
- To zoom in or out, use your scroll wheel, or the button along the bottom of the preview pane.

COMMENTING OUT OR EDITING PARTS OF A MODEL

We have put explanations at key points in our models to tell you what is going on. These notes are called *comments*. There are two ways to do this, and you will see comments at key points in the models in the repository to help you make changes if you want to. The first way to comment out part of a line is to put two backslashes like this ahead of what you want to ignore.

```
a = 5 ; //This is the value of a
```

Or, you can enclose several lines in **/*** and ***/**, as follows, and OpenSCAD will ignore all of it.

```
/*
This is a block of comments
Blah blah blah
*/
```

In some cases, we have included options in the code, with some of them "commented out." You would remove the // from in front of the version you want. For instance, the model might say:

```
//a = 5; // use for a small model
//a = 20; // use for a larger model
```

If you wanted the larger model, you would take away the **//** ahead of the **a = 20;** to get:

```
//a = 5; // use for a small model
a = 20; // use for a larger model
```

If you are new to computer code at this level, you might find a book or website about the basics of programming in the C, C++, or Java language helpful. For now, you can just copy the one- or two-line examples in Chapter 3 and beyond in the book, or the models from the repository, and use them as-is or just change a number here and there. If you want to go farther, though, obviously you can keep going. The "Programming Tips for OpenSCAD" section later in this chapter is intended to help with that (but presumes you have done some computer programming already).

Some of the chapters in this book suggest you change the models in various ways. In some cases, you will just change a number, preview it, and then render and export to STL as described in the "OpenSCAD Workflow" section of this chapter. In other cases, you may need to comment out one line and remove the "//" before another to select a different option. Each chapter has detailed directions on how to do that for the model it uses.

CREATING BASIC SHAPES

OpenSCAD allows you to create complex 2D and 3D shapes by combining basic shapes, which we will call *primitives*. The basic 2D shapes available in OpenSCAD are a circle, square, and polygon (a closed 2D curve with an arbitrary number of straight sides). In 3D, the language gives you a sphere, cube, cylinder, and polyhedron (which lets you define arbitrary 3D shapes). That may not sound like a lot, but with some creative tricks you can generate just about any shape from these.

Let's look at some simple examples; more complex examples are the subject of Chapter 3. To create a 2D square 5mm on a side, you would type this line:

```
square(5);
```

You can't 3D print a square, but you can use a command to extrude it into the third dimension. To create a cube in OpenSCAD that is 5mm on a side, you would write:

```
cube(5);
```

Notice that the 5 is enclosed in a pair of parentheses and that the line ends in a semicolon. Every object in OpenSCAD ends in a semicolon. Suppose, though, that you don't want to print a cube, but rather want to print out a rectangular solid that is 5mm in one dimension, 10mm in the other, and 15mm in the third. You would write:

```
cube([5, 10, 15]);
```

Notice that the inner brackets are square. And yes, even though strictly speaking this is no longer a cube, OpenSCAD still calls it one!

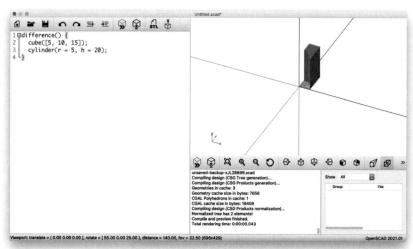

FIGURE 2-5: Demonstration of the difference() function, where we are subtracting a cylinder from a cube.

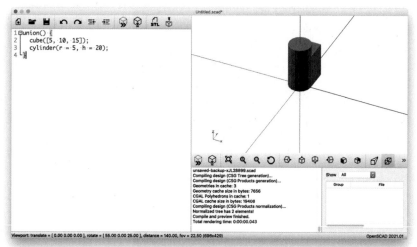

FIGURE 2-6: Demonstration of the union() function.

Cylinders need two numbers specified, the radius (half the distance across the top or base) and height. A cylinder with a radius of 10mm and height of 20mm would be:

```
cylinder(r = 5, h = 20);
```

COMBINING BASIC SHAPES

When we want to make more complicated shapes out of simple ones, we want to be able to add them together, subtract them from each other, and

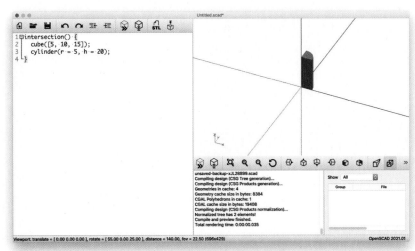

FIGURE 2-7: Demonstration of the intersection() function.

maybe see where they occupy common space. These operations are known as *Booleans*, in honor of British mathematician and philosopher George Boole who first codified them in the mid-1800s.

OpenSCAD lets you combine shapes by subtracting them using the **difference()** function (Figure 2-5); by adding them (Figure 2-6) using **union()**; and lets you select where two shapes overlap (Figure 2-7) by using **intersection()**.

The OpenSCAD format of these Boolean functions is to list the function, and then enclose the two functions you are subtracting, combining, or intersecting from each other in a set of curly brackets. The order matters for **difference()**, but not for **union()** or **intersection()**.

COORDINATE SYSTEMS IN OPENSCAD

You may have noticed that we have been talking about 3D objects, and not coincidentally we are providing OpenSCAD with three dimensions for some of these objects. To explain how this works, we need to talk about *coordinate systems*. The commonest one is called the *Cartesian* coordinate system, after René Descartes, a French mathematician who lived in the early 1600s. He came up with the idea of graphs as well as many of the ways we write equations and think about them. Legend has it that he got the idea for Cartesian coordinates lying in bed and watching a fly walk around the tiled ceiling, or, in some versions, on a window with small panes.

In any case, you can see Descartes' heritage in OpenSCAD. If you look

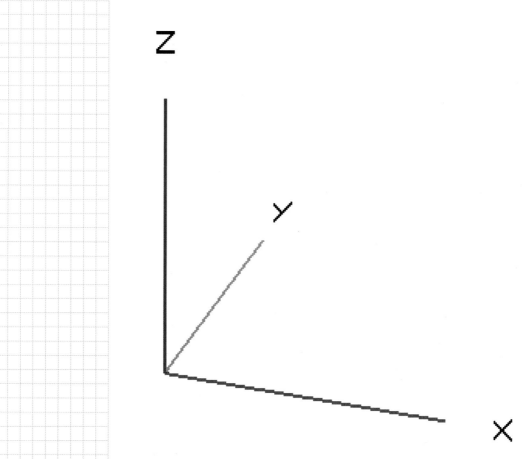

FIGURE 2-8: The coordinate system in OpenSCAD

closely, you will see a small symbol in the lower-left corner of the display window that is marked with an x, y, and z (Figure 2-8). These are the three directions in which we will measure how big the objects we are creating are, or how far we have moved them. These directions are called *axes*, and three of them, to each other, are the *Cartesian coordinate system*.

As you can see if you move the view around a bit in OpenSCAD, any pair of them make a corner of a square. The axes only all cross at one point called the *origin*. There, by convention, x = 0, y = 0, and z = 0. OpenSCAD shows the axes ending there, but in reality you can go the other direction, too, where the values of x, y, or z are negative.

3D printers typically use Cartesian coordinates too and usually consider x and y to be the plane of the 3D printer's platform, and z to be the vertical axis.

By default, when you bring a model into your slicer, it will be moved so that it is centered on the platform, with its lowest point at z = 0.

We will follow that convention in this book. This is also called a right-handed coordinate system, because if you imagine taking your right hand and resting the outside of it on the x-axis, if you curl your fingers toward the y axis your thumb is pointing upward, in the z-axis direction.

If you open up OpenSCAD and type this line:

```
cube([5, 10, 15]);
```

You will indeed discover that the cube is 5mm long in the dimension lined up with the x-axis, 10mm long in the y-axis, and 15 in z.

In Chapter 3, we will learn about how to fly around in the Cartesian system, and learn basic geometry manipulations. First, though, we highly recommend that you play with the simple ideas in this chapter and do some experiments. OpenSCAD also has a lot of examples in the program, available at **File > Examples**.

Lines that define an object need to end in a semicolon. The OpenSCAD manual (available under the "Documentation" tab at openscad.org) has a very good reference for the details of exactly when you need a square bracket, curly bracket, and so on. In Chapter 3 we walk through examples of how to build up (and explore) objects.

PROGRAMMING TIPS FOR OPENSCAD

If you are already a programmer, here are a few things to know about OpenSCAD that are a little idiosyncratic and different from the languages it otherwise resembles. We don't go into the things that are pretty much the same as other languages in the C family (C, C++, Java, etc.).

NO TRUE VARIABLES

Most programming languages allow you to do something like this:

```
a = 5;
```

Computers, of course, are working in ones and zeroes, and code like this looks scary at first. But let's imagine that we are the computer executing this code. We might look at this line and say, "I am going to take a sticky note and label it **a**, and then write the number 5 on it." Programmers would call this sticky note a *variable*. This type of code is common in languages like C. OpenSCAD is fine with this.

You can also do some math to assign a value to a variable. For example, you could say:

```
a = 5;
b = a + 1;
```

In this case, you would end up with two variables, one called **a** and one called **b**. The variable **a** would store the number 5, and **b** would store the number 6. OpenSCAD is fine with this, too.

However, in most programming languages you can also do something like the following:

```
a = 5;
a = 6;
```

In this case, you are again starting by setting **a** to 5, and you might have some code in between these lines that uses the variable with its value of 5 before you get to the next line. There, you change the value to 6, essentially crossing out the number on your sticky note and writing in a new one.

Variables don't work that way in OpenSCAD, though. OpenSCAD will go through your code and figure out what all of the variables are before it does anything else. In this example, the variable **a** would be 6 for the entire time (even before the line where you set that value). This also means that you can't do things (common in other languages) like:

```
a = 5;
a = a + 1;
```

This won't work at all. OpenSCAD will produce a warning and leave the variable **a** with an undefined value. This is because when you try to assign a value to **a** on the second line, it replaces the first value assignment instead of modifying it, so you end up with only the second line. When it tries to read that line, it has a variable called **a** that hasn't been defined yet, so you can't add one to that value and store it.

The bottom line is that OpenSCAD lets you set a variable to a value once, and then that's it. If you are familiar with the C programming language, it is sort of like a constant (**const**) in C. For the nuances of this, see the User Manual on the OpenSCAD site.

OPENSCAD IS CASE SENSITIVE

The OpenSCAD language is case-sensitive, so **Cube()** or **CUBE()** will not do anything, and **apple** is different from **Apple** or **APPLE.** In the fragment below, we define two separate variables with different names (one lower case, one all caps). The program will print **"apple", 1** in the console window, and ignore APPLE. It is often useful to have the program tell you what is going on; the "echo" command will do that.

```
apple = 1;
APPLE = 2;
echo("apple", apple);
```

IF AND ELSE

Very often we want to make a decision in our code, to do one thing if a number is greater than some limit, for example, and to do something else otherwise. We can do that by commenting out the code, as we noted earlier. Or we can use an "if" statement. This code will set the variable **a** to be equal to 0.05, and then checks to see if it is greater than 5. It is not, so we jump to the next line and create a sphere of radius 5. This is a way to make sure you aren't trying to create something too small to print, for example.

```
a = 0.05;
if (a > 5) sphere(a);
else sphere(5);
```

Then you could just change the value of **a** to whatever you wanted, and the code would make sure that you didn't try to print a sphere that was too tiny.

VECTORS AND RANGES

In addition to single numbers, OpenSCAD also lets you create a variable that refers to a list or sequence of numbers. These are called *vectors* and *ranges*. You've already seen one example of a vector above when we made a "cube" with different dimensions in the x, y, and z directions. It's common to see a comma-separated list of x, y, and z values (in that order) in OpenSCAD, enclosed in square brackets.

```
point = [x, y, z];
```

Vectors don't always have to be 3 elements long, though. You can make vectors (commonly called arrays in other programming languages) of any length you want. This is useful for storing a list or sequence of values, like this:

```
fibonacci = [1, 1, 2, 3, 5, 8, 13, 21];
```

You can read a specific value from a vector using its *index*, which is the number of its place in the vector. These are zero-indexed, which means that the first value will have an index of zero, and the last value will have an index of one less than the total number of values in the vector. Using the **fibonacci** list we just used as an example, we would print 1, 21 if we used this statement:

```
echo(fibonacci[0], fibonacci[7]);
```

If you have a sequence that just adds a fixed amount at each step, you can use a *range* instead. A range looks similar to a vector, but it can only have two or three values, and those values are separated by colons instead of commas. The first value tells you where to start counting, and the last value tells you where to stop. In a range with three values, the middle one is a "step" value, which tells you how much to add each time.

```
range = [start:end];
range = [start:step:end];
```

If you only have two values, the step value will be 1. (We have shown the values that each range produces as a comment after each statement.)

```
a = [0:5]; // 0, 1, 2, 3, 4, 5
b = [0:5:20]; // 0, 5, 10, 15, 20
c = [2:5:20]; // 2, 7, 12, 17
d = [10:-2.5:0]; // 10, 7.5, 5, 2.5, 0
```

Note that ranges will be counted up to and including the end value, but the end value will not be included in the range if it is not equal to the start value plus some multiple of the step value. Ranges also don't let you, for example, pick out the tenth value in the sequence as vectors do. What use are they then? Both vectors and ranges can be used to control a loop.

LOOPS

Very often in writing computer code we want to do something over and over, often done with a structure called a *loop*. A loop iterates over the values in a vector, or the ones produced by a range. It looks like this:

```
for(i = [0:1:5]) {
        echo(i);
}
```

A loop will run whatever's inside it for each value produced by the vector or range, temporarily treating the variable (in this case, **i**) as if it had that value. This example would output six separate lines in the console, each with a different number going from 0 to 5. The number in the middle is optional. If you instead wrote **[0:5]**, OpenSCAD would assume you wanted to increment the value by 1 each time, and you would get the same result.

FUNCTIONS

Functions in OpenSCAD are more like mathematical functions than functions in traditional computer languages. That is, a function can just return one thing (which can be an array). So:

```
x = 1;
y = 2;
z = 3;
a = myfunction(x, y, z);
function myfunction(apple, banana, pear) = apple *
banana * pear;
echo (a);
```

Which will print a 6, or 1 times 2 times 3. (OpenSCAD uses the convention of an asterisk for multiplication.)

Functions are allowed to call themselves (what a mathematician would call recursive) as long as you have a way to tell it to stop! The classic example of this is the *factorial* function. A factorial is a number times itself minus 1, times itself minus 2, and so on. For example, 4 factorial (written 4!) is:

$$4! = 4 * 3 * 2 * 1 = 24$$

In OpenSCAD, this would be:

```
function factorial(n) = n ? n * factorial(n - 1) : 1;
```

OpenSCAD does make available a lot of the standard math functions automatically (you don't have to do anything fancy to include libraries). Again, there's a good list in the manual. Finally, unlike most programming languages, angles in OpenSCAD are assumed to be in degrees, not radians.

OPENSCAD VARIANTS

Since OpenSCAD is a freely available program, there are various other systems built on it. Some of our models push the boundaries of OpenSCAD and may not work with these programs, depending on what they have altered. We have stuck with the core version of OpenSCAD which has the virtues of being well-documented, open source, and free. The models here were developed and tested with OpenSCAD version 2021.01.

There are a few emerging drag-and-drop block-coding environments based on OpenSCAD, and you may want to keep an eye on those to see if they stay functionally close enough. However, some of these are based on OpenJSCAD, which is a reworking of OpenSCAD into Javascript. OpenJSCAD code has a different syntax than OpenSCAD and some different primitives, so the models will not translate without significant rework.

TINKERCAD

Tinkercad is a drag-and-drop simple 3D CAD program. The great virtue of OpenSCAD is that everything is drag-and-drop. This means you can grab a geometric shape, pull it onto a virtual build platform, and alter it in various ways. Tinkercad is available for free at **www.tinkercad.com**, provided by Autodesk. Figure 2-9 shows the workspace. Parts can be added, subtracted, rotated, and scaled up or down by dragging on the screen. Tinkercad runs in a browser window, so there is nothing to download and it will work on Chromebooks (relatively rare among CAD programs).

The negatives are similar to the positives, in that it is challenging to be very precise with Tinkercad. However, in Chapter 3 we will discuss how you might learn some of the basic geometry manipulations in Tinkercad as well as OpenSCAD. Finally, there is some ability to use drag-and-drop code to assemble objects (Figure 2-10). There isn't enough there as of this writing to do many of the models in this book, but it is something to watch for the future.

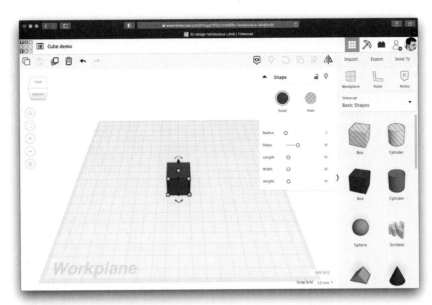

FIGURE 2-9: The Tinkercad workspace

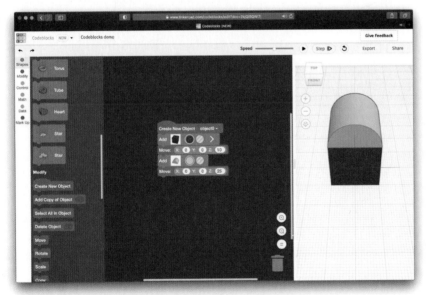

FIGURE 2-10: Tinkercad block coding interface.

SUMMARY AND LEARNING MORE

In this chapter, we learned the basics of OpenSCAD. We also saw where to download the models for this book on Github and how to open those models in OpenSCAD. We also covered a bit about 3D printers and materials.

If you want to know more about OpenSCAD, as we have noted in the chapter there is a very good manual at openscad.org. There are also tutorial videos linked on the website. We are particularly partial to the "cheat sheet" that lists examples of all the different OpenSCAD features on one page. You should be able to do the projects we describe in this book purely by making small (or no) changes to models we have built already and put in the repository. However, to go further some coding knowledge would be helpful, and we suggest you find some tutorial videos about the C programming language. Tinkercad also has tutorials linked on its site, if you want to experiment with it as well.

In the next chapter, we will go into more features of OpenSCAD (and a little bit about Tinkercad alternatives for some activities). In particular, we will see how OpenSCAD can be used as a geometry simulator, trying out geometrical concepts by changing a line or two in a model.

CHAPTER 3
SIMULATING
GEOMETRY

In Chapter 2, we introduced the basic mechanics of OpenSCAD. In this chapter, we will first introduce some concepts you'll need as we start talking about more complicated shapes. Then, we'll introduce our first real model, and talk through what it does and how to modify it.

Then, we'll jump into using OpenSCAD to simulate some of the basic ideas in geometry. If you were to compare the table of contents of a geometry book and the list of commands in OpenSCAD, you would notice a lot of overlap. This is, after all, why this category of CAD programs is called "Constructive Solid Geometry."

We highly recommend that as we walk through the chapter, you actually type the short models into OpenSCAD as we discuss them. As you do this, you should mess with the numbers, change things, and see what happens.

In this chapter, we'll also introduce a few of the more sophisticated coding constructs and OpenSCAD features that we use in upcoming models. If you aren't interested in learning to make models yourself you can skip some of this detail, but we hope that you are inspired to take the program out for a spin too. In some cases, you'll need material from later in the book to fully appreciate what is going on. In those cases, we will note where the material is explored later in the book, but introduce the OpenSCAD capability here. That way, the OpenSCAD-focused (versus geometry-focused) material is in one place for you as a reference.

We will walk you through using one of the models to create Platonic solids, which are objects whose faces are all regular polygons, all of which meet at equal angles. This will give you some practice with using a model from the repository and modifying it.

We'll guide you through a model that makes a simple (but very extensible) castle, to tie together OpenSCAD and geometrical concepts and give you some experience with going step-by-step through a model. Finally, we'll also briefly discuss Tinkercad, a simpler CAD program that can do some of this simulation as well.

3D Printable Models Used in this Chapter

The following models are discussed in this chapter. See Chapter 2 for directions on how to download them. There are also quite a few model fragments scattered in this chapter. Be sure to type them in and try them out.

platonicSolids.scad
Prints one of each of the five Platonic solids. This model scales all five so they are the same height as each other on the print bed.

edge_platonic_solids.scad
Prints one or more Platonic solids, scaled by the user input length of one edge.

castle.scad
Prints a castle, developed as an exercise in the various moving and scaling functions in this chapter.

PRELIMINARIES

Before we start typing some more interesting models in OpenSCAD, we need to get through a few definitions and background concepts. You might have used some of the words before, but we'll use them a little more specifically here than you might in everyday discussion. Mathematicians can be awfully picky about exactly what a word means, so we'll give you their definitions before we start hanging out with them.

CIRCLES AND PI

A *circle* is a closed curve that is a constant distance, the *radius*, from its center. You've probably heard that the area of a circle is pi times the radius squared, but what is pi? It is written as the Greek letter π, and pronounced like the word "pie". It is the ratio between the circumference, the distance one would travel to go all the way around a circle, and the *diameter*, which is twice the radius, or the longest distance across a circle.

Pi also appears in many other branches of math, and is an *irrational number* that equals roughly 3.14159. Irrational doesn't mean the mathematicians who came up with pi are crazy. It means that there is no way to show pi as a fraction — a *ratio* — of one whole number over another, as you can with values like 10/3 and 3/5. (Numbers like 3 are rational because you could express them as 3/1, since anything divided by 1 is just the number itself.) You can roughly approximate pi with 22/7 or (slightly closer) with 355/113, but these are not exact.

You might think 10/3 is irrational since it is a decimal that never ends. It's 3.333... on to infinity. The difference is that the fractional part *repeats* (or terminates) for rational numbers, but does not in an irrational one. For example, 10.3 and 10.4 are rational numbers, since they could be written 103/10 and 104/10, respectively.

People have computed pi to tens of trillions of digits, and there's never any end in sight. (That's not a typo. Tens of trillions, really.) You can see the first million at piday.org. If you really want to get into it, Pi Day is March 14 (3.14), and geeky observances abound around the world, usually involving eating pizza or other kinds of pie in the pun-loving parts of the English-speaking world. For some years, MIT released its offer of admissions on March 14, and took it some more digits in 2020 by moving the release to 1:59 PM (3.14159...).

There is also another symbol for 2π—the Greek letter tau, τ. Some people think tau is more intuitive when talking about circles than pi is, but it never really has caught on as an alternative. Pi comes up all over the place in geometry, and we needed to let you know about it before we dive too far into other shapes. OpenSCAD bases a lot of other shapes on the circle (or cylinder) as we will see a little later in this chapter.

ANGLES AND DEGREES

Stand in one place facing forward and stick your right hand at shoulder height directly in front of you. Now swing your right arm around clockwise until your hand is pointing directly to the right. You just swept your arm through an angle of 90 *degrees*. A degree in geometry is a measure of how much something has rotated about its center. 90° angles are often called "right angles." Wikipedia says this is from the Latin word *rectus*, for "upright," since something sticking straight up forms a 90° angle from the ground.

If you sweep through 90 degrees (written 90°) and keep going for 90 more, you will be pointing in the exact opposite direction from where you started. Two times 90° is 180°, so you can think of a straight line as a 180° angle. If you kept going all the way around to bring your hand back to where you started, you would have turned 360°, which is a full circle.

PARALLEL AND PERPENDICULAR

Imagine you have a perfectly rectangular box. The top and bottom of the box are always at the same distance from each other everywhere if the box was perfectly square at all its corners. Now imagine that the top and bottom of the box went on to infinity; that would still be true.

Planes (or lines) that never meet (another way of saying they stay the same distance from each other everywhere) are said to be *parallel* to each other. An example of a pair of parallel lines would be two opposite edges of the bottom of a (finite) box. If you drew a line through each and extended those lines to infinity, they would never cross.

Where two sides of the box meet, however, they intersect at a right angle. Surfaces (or lines) that intersect at 90 degrees are said to be *perpendicular* to each other. You can explore this with a box, or print out a cube in OpenSCAD with this command, which prints out a cube 40 mm on a side.

```
cube(40);
```

The top and bottom sides of the cube are parallel to each other. The top or bottom and any side are perpendicular. Opposite sides of the cube are parallel to each other.

CODE VERSUS MATH CONVENTIONS

Some things are a little different in OpenSCAD and regular math. Let's talk about a few of those here.

EQUAL SIGN

When we see 2 + 2 = 4 in a math class, we know that anything we do on one side of the equal sign we can do on the other. For instance,

$$2 + 2 + 2 = 4 + 2$$

In a math class, the equals sign says, "the stuff on either side has to be the same." However, in a computer program, an equals sign sort of means "becomes." So we might say:

```
apples = 2;
```

That means that wherever the word **apples** appears in the code, you can replace it with a 2. (In OpenSCAD at least — this is different in different programming languages.) But you can't say:

```
apples + 2 = 2 + 2;
```

without the program giving you an error message. But you could say:

```
apples = 2 + 2;
```

In which case **apples** would be 4 after that line of code was executed by the computer. You also can't say:

```
2 = apples;
```

OpenSCAD would think **2** was the variable, and you'd be trying to set it to the value of a variable you haven't defined called **apples**. Unlike algebra, the equals sign in OpenSCAD has a direction to it.

DEFAULTS

We will start to use more complicated OpenSCAD functions in this chapter.

Most of them have a bunch of variables listed, but in our examples we will only use a few. The rest of the variables are assumed to have *default* values. In the OpenSCAD manual, you might see more than we even list since functionality gets added to the program all the time.

Default, in this case, means that the program says, "unless you tell me otherwise, I'm assuming that all the other variables have certain predefined values." For example, most OpenSCAD shapes have a variable called **center,** which is defaulted to **false**. The meaning varies a bit depending on the shape (see the OpenSCAD manual), but for example **center = false** for a cylinder creates the cylinder upward from the x-y plane, and **center = true** has half of the object above and half below the x-y plane.

Your math teacher probably won't let you get away with saying, "just assume all the default values" but OpenSCAD will, with a few exceptions. You usually have to supply dimensions, for example. A cube 4mm on a side in OpenSCAD is:

```
cube(4);
```

But, as we saw in Chapter 2, we can have a "cube" (in OpenSCAD's opinion) with different-length sides. If we only give **cube()** one number, the default is that all of the sides are the same. Therefore the last example can also be written as:

```
cube([4, 4, 4], center = false);
```

which would be pretty annoying to type over and over.

POLYGONS AND POLYHEDRONS

Now that we have a few definitions out of the way, let's meet some new shapes and see what they look like in OpenSCAD. *Polygons* are 2-dimensional mathematical shapes that have at least three sides, and that completely split a space into an area outside the polygon and one inside it. There are special cases that have been known and studied for a few thousand years, and they form the basis of much of the geometry we know today. We'll take a look at those first.

REGULAR POLYGONS

Polygons that have all edges the same length and all the angles between the edges the same are called *regular polygons*. Where the sides connect, they

FIGURE 3-1: A regular pentagon, hexagon, and octagon

form a *vertex* (plural, *vertices*). An example is the corner of a square. Some of the regular polygons are a triangle with all sides equal (3 sides), a square (4 sides), and the regular pentagon (5 sides), hexagon (6 sides), and octagon (8 sides). Each of these shapes, except for the square which is defined to have all its sides the same, can have sides of different lengths, but then they would no longer be regular. We'll spend all of Chapter 5 exploring triangles in-depth, and then see in later chapters how many other polygons you can build up from them.

OpenSCAD can easily create regular 2D polygons, and then extrude them in the third dimension to be 3D-printable. The following single line in OpenSCAD will print a pentagon, extruded to be 10mm tall:

```
linear_extrude(10) circle(r = 15, $fn = 5);
```

Making **$fn** equal to 5, 6, and 8 will give the results shown in Figure 3-1

You might wonder why we are using the "circle" function to create a pentagonal solid and why the pentagon is not 15 mm on a side. Let's deconstruct what is happening here, since it will give you insights into how the program thinks.

The variable "r" is called the *radius of the polygon*; it is the radius of the circle that just touches each vertex, or the *circumscribed circle*. This is the smallest circle that the polygon will fit inside, shown in blue (See Figure 3-2). Some books call this the *circumradius* of the polygon, but we will use the simpler convention of just calling it "the radius of the polygon." The circle that just touches the center of each side of the polygon, is called the *inscribed circle*. It

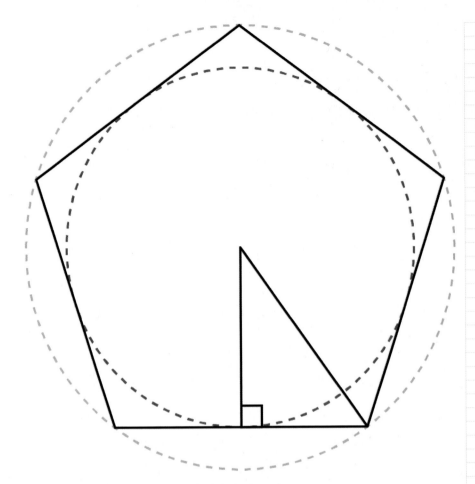

FIGURE 3-2: Inscribed and circumscribed circles in a polygon

is the largest circle that will fit inside the polygon, shown in red. This works both ways: the polygon is *inscribed* in the blue circle (see chapter 7 for more). The red circle's radius is the vertical line, and is called the *apothem* of the polygon.

OpenSCAD creates a circle by using short line segments, which it calls *fragments*. The **$fn** variable is the number of fragments used to create a closed regular polygon. Thus **$fn = 5** makes a closed curve consisting of five equal straight lines with equal angles between them, otherwise known as a regular pentagon. If you used **$fn = 4**, you would get a square. **$fn = 6** is a hexagon, and **$fn = 8** is an octagon. As you use a bigger and bigger value of **$fn**, you'll see that the polygon looks more and more like a circle. We will explore this in Chapter 7, when we talk about inscribed and circumscribed polygons. You can generate a pretty good circle with

```
linear_extrude(10) circle(r = 15, $fn = 100);
```

Incidentally, be careful of setting **$fn** too high on complex shapes. This can make OpenSCAD take a long time to process, or even crash (and lose all your model coding). Be sure to save your model before trying to preview or render it.

POLYHEDRONS

Now that we have looked at 2D shapes, we can move up to 3D. Polyhedrons are solid objects that do not have any curved surfaces. For example, a 3D rectangular box is a polyhedron, but a cylinder is not (because its sides are curved). Each side of a polyhedron (like the top or bottom of a cube) is called a *face*, and we usually call the line connecting two faces an *edge*.

Just as there is a special category of polygons with all sides equal, there is also a category for solids that have the same regular polygon making up all their sides. These are called *Platonic Solids*, and there are only five of them.

PLATONIC SOLIDS

There are just five solids that meet the definition of Platonic solids, named after the Greek philosopher Plato, who lived about 2400 years ago and was very into things being symmetrical and regular. To be a Platonic solid, the edges must be equal, all the angles at which those edges meet must be equal, and the same number of faces must meet at each vertex. They are the tetrahedron (4 triangular faces), the cube (6 square faces), the octahedron (8 triangular faces), the dodecahedron (12 pentagonal faces), and the icosahedron (20 triangular faces). We will see a lot more of them in Chapter 9.

PLATONIC SOLIDS MODELS

The file **platonicSolids.scad** can be used to create these objects, as shown in Figure 3-3. It creates all of the solids at once, scaled to be **size** millimeters tall when they are resting on one face. If you elect to scale the models in your slicing software, note that they will cease to be Platonic solids if you do not scale them uniformly. You can however change the value of **size** and all the solids will scale consistently.

We also have a second file, **edge_platonic_solids.scad** which allows you to print a set of Platonic solids defined by the length of each edge of the polygons. You might find this one more useful later on in Chapter 9 when we study the surface area, and it is more convenient in general for calculation to

FIGURE 3-3: The Platonic solids

specify it this way. In this case, you would make the variable **edge** whatever length (in millimeters) you wanted, within reason. (You would, of course, need to make it small enough so that it fits on your printer.)

You might notice in that model that after defining the variable **edge** there is a block like this:

```
translate(edge * [.75, -.75, 0]) tetrahedron(edge);
translate(edge * [1.5, 1.5, 0]) cube(edge);
octahedron(edge);
translate(edge * [2, 0, 0]) dodecahedron(edge);
translate(edge * [.5, 1.6, 0]) icosahedron(edge);
```

If you wanted to skip creating any of these models, you could comment out ones you did not want. (See Chapter 2's discussion on altering models.) For example, this would 3D print the cube and octahedron, and ignore the others (which are commented out with "//" at the beginning of the line):

```
// translate(edge * [.75, -.75, 0]) tetrahedron(edge);
translate(edge * [1.5, 1.5, 0]) cube(edge);
octahedron(edge);
// translate(edge * [2, 0, 0]) dodecahedron(edge);
// translate(edge * [.5, 1.6, 0]) icosahedron(edge);
```

You will observe that the tetrahedron, octahedron, and icosahedron are made up of triangles, the cube is made of squares, and the dodecahedron is made of pentagons. The reason there are only five Platonic solids comes from the equal-angle constraint. If you think about it, the angles at each

FIGURE 3-4: A tetrahedron and double tetrahedron

vertex have to be less than 360 degrees. (Imagine the vertex flattened out; at most, it can be a full circle.) At least three sides have to come together at each vertex, so that means that the angles of each side must be less than 120 degrees. This means that only triangles, squares, and pentagons work for the faces. If you work out the combinations, it turns out that the ones here are the only options. Platonic solids are often used for dice because if their weight distribution is consistent, the die will be fair.

EULER'S CHARACTERISTIC

The Swiss mathematician Leonhard Euler lived in the 1700s. He took much of the Greek geometry and expanded and modernized it. One of the key things he realized was that there was a fundamental relationship between the number of vertices, edges, and faces in any convex polyhedron, regular or not. A convex polyhedron is defined as one that, if you drew a line from any vertex to any other one, the line would be entirely contained inside. In other words, you would not have to cross an air gap to connect any two vertices. The relationship is known as *Euler's Formula* or, more commonly, *Euler's Characteristic* (because Euler came up with a lot of formulas):

Vertices – Edges + Faces = 2

Adding one vertex to a tetrahedron requires adding 3 edges and 3 new faces, which collectively would replace one of the original faces. This would leave you with something that looked like two tetrahedrons glued together. If you want to test this out, make two tetrahedrons and line up one face of each. That will give you a double tetrahedron (Figure 3-4).

> Tetrahedron: 4 vertices, 6 edges, 4 faces
> 4 - 6 + 4 = 2
> Double tetrahedron: 5 vertices, 9 edges, 6 faces
> 5 - 9 + 6 = 2

TRANSLATION

In the last chapter, we saw how to create various solids. However, now we want to move and reorient these objects so that we can build them up into more interesting shapes. OpenSCAD has a variety of what it calls *transformations*. We will look at a few basic ones in this section, and then some of the more sophisticated ones later in the chapter after we introduce a few more ideas.

As you build up more complicated models from simple pieces, you'll want to move around the pieces in various ways. If you just want to move an object, you would use **translate([x,y,z])**. Note that OpenSCAD uses parentheses for all its built-in modules, and square braces to contain sets of numbers. In the case of **translate(...)** the three numbers stand for the distance, in millimeters, you want to move the object. This line:

```
translate([0, 0, 5]) cube(10);
```

creates a cube 10mm on each edge, and would move it up 5mm in the z (vertical) direction. Note that there is a semicolon after the object (**cube**) but not after the **translate(...)**. You could think of the semicolon as the period at the end of a sentence. It ends the definition of one object and lets you start another. If you forget a semicolon, you will usually get an error. OpenSCAD often displays that error on the following line, because the problem occurs when you try to start a new "sentence" without ending the previous one.

OPENSCAD ORDER OF OPERATIONS

OpenSCAD starts reading each line from the object (like, say, **cube(10)**) and then applies transformations from nearest the object to farthest from it, from *right to left*. So in the case of the previous example, it will make a cube and then translate it. We will see that this matters, in upcoming examples involving multiple rotations of an object, or both moving an object and rotating it. Each object definition will normally end in a semicolon, so you start at the semicolon and work your way to the left, defining the type of object first, then performing whatever operations like translation, rotation (next section!), and so on in that order.

You might be asking why the creators of OpenSCAD would make you read the code from right to left. Technically, what you're doing is reading from the

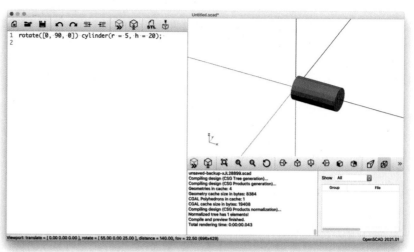

FIGURE 3-5: The cylinder rotated 90 degrees about the y axis

inside out. In the line of code above, the cube module is *enclosed* in the translate module. The same code could also be written, a bit less succinctly, as:

```
translate([0, 0, 5]) {
  cube(10);
}
```

This convention makes it possible to enclose multiple objects, so that you can apply transformations to them as a group. For example, this would translate two cubes at the same time:

```
translate([0, 0, 5]) {
  cube([20, 10, 10]);
  cube([10, 20, 10]);
}
```

ROTATION AND MIRRORING

The transformation **rotate([x, y, z])**rotates the object around the x, y or z axes. The values are the angle, in degrees, to rotate about that axis. For example,

```
rotate([0, 90, 0]) cylinder(r = 5, h = 20);
```

would result in a cylinder on its side, as shown in Figure 3-5.

We find it easier and more predictable to rotate one axis separately at a time. So, if we were to rotate a cylinder about x and then y axes by 30 degrees each, we would write:

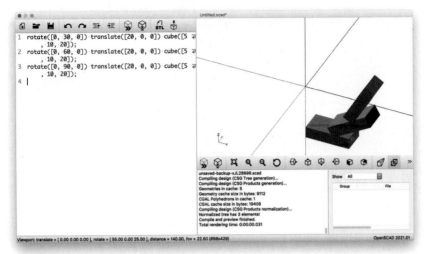

FIGURE 3-6: Translating and then rotating

```
rotate([0, 30, 0]) rotate([30, 0, 0]) cylinder(r = 5, h =
20);
```

OpenSCAD supports rotating about more than one axis at once, and will rotate about x first, then y, then z. When you start to combine more than one transformation (be it multiple rotations, or a mix of translation and rotation) the order of operations matters. If you have translated an object first, the object will rotate around the axis and maintain whatever the translation was. If you translate 20mm in x, and then rotate about the y axis, for example, to create three rectangular solids ("cubes" with different size edges) like this:

```
rotate([0, 30, 0]) translate([20, 0, 0]) cube([5, 10, 20]);
rotate([0, 60, 0]) translate([20, 0, 0]) cube([5, 10, 20]);
rotate([0, 90, 0]) translate([20, 0, 0]) cube([5, 10, 20]);
```

The resulting three pieces are shown in Figure 3-6. You can see that the rectangular solid can be thought of as sliding around a cylinder of radius 20 mm that is centered on the y axis.

However, if instead you were to try (rotate and then translate) you would get the result in Figure 3-7.

```
translate([20, 0, 0]) rotate([0, 30, 0]) cube([5, 10, 20]);
translate([20, 0, 0]) rotate([0, 60, 0]) cube([5, 10, 20]);
translate([20, 0, 0]) rotate([0, 90, 0]) cube([5, 10, 20]);
```

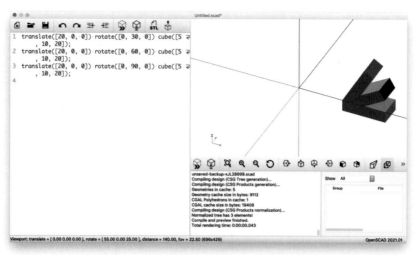

FIGURE 3-7: Rotation and then translation

TRANSLATING OR ROTATING MULTIPLE OBJECTS

The order of rotation and other transformations matters. They are read from the inside out, closest to the object first. You can think of these functions as adjectives, and the shapes as nouns. These transformations don't do anything on their own, if there is no associated shape. If you want to translate or rotate multiple things at once, you can enclose them in curly braces. For example, the previous example could equivalently be written:

```
translate([20, 0, 0]) {
  rotate([0, 30, 0]) cube([5, 10, 20]);
  rotate([0, 60, 0]) cube([5, 10, 20]);
  rotate([0, 90, 0]) cube([5, 10, 20]);
}
```

This also has the virtue that it makes the order in which we are rotating and then translating more explicit.

MIRRORING

Finally, `mirror([x, y, z])` will flip (mirror) an object. This is done by putting a zero in the coordinates you do not want to flip, and a 1 in the coordinate you do want to flip.

Therefore, `mirror([1, 0, 0])` flips an object along the x-axis, `mirror([0, 1, 0])` along the y-axis, and `mirror([0, 0, 1])` along the z-axis. This line takes a rectangular solid that is 5mm by 10 by 20, rotates it 30 degrees

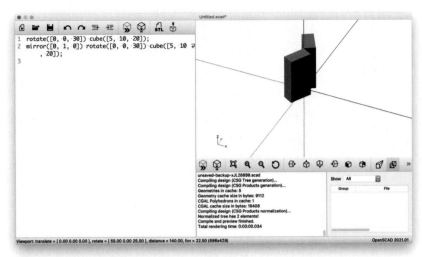

FIGURE 3-8: Before and after mirroring a rotated cube

around the vertical axis, and then flips it to the other side of the vertical axis. Type this line into OpenSCAD and try it out to get some intuition (Figure 3-8).

```
mirror([0, 1, 0]) rotate([0, 0, 30]) cube([5, 10, 20]);
```

Since all these transformations are easier to see when you change values and see what happens, we highly recommend you play with these functions a little before reading on and get some intuition by seeing how they work by messing with some examples. You should generally only try to mirror along one axis at a time, because values other than those shown can produce unexpected results.

SCALING

A different type of transformation is making your objects bigger or smaller. Of course, you can scale all OpenSCAD objects just by using bigger or smaller dimensions when you create them. But if, for instance, you want to start with one shape and progressively add copies to a project that gets progressively bigger, you might want to use the **scale([x, y, z])** transformation. In this case, the x, y, and z variables are the factor you want to scale your existing object by. If, for example, you had a cube 10mm on a side and wanted it to be 20mm instead in the x direction only, you would use this line of OpenSCAD:

```
scale([2, 1, 1]) cube(10);
```

Note that you need to include a scaling factor of 1 for the axes you want to leave alone. If you try to scale any axis by 0, the object will vanish, since one of the dimensions is now zero thickness!

MATH FUNCTIONS

Sometimes you want to do some math when you are creating your models. OpenSCAD allows you to build in many things we will learn about in later chapters. There's a complete list in the OpenSCAD manual (under the "Documentation" tab at OpenSCAD.org.) For example, to multiply a number by itself (raise it to a power) you need to use the **pow(a, b)** function. As of the 2021.01 release of OpenSCAD, you can also raise to a power with the "^" operator. If we wanted to make a cube that was 3 * 3 = 9mm on a side, we could write:

> **cube(pow(3, 2)));**

> Or, alternatively, (in OpenSCAD 2021.01 or later):

> **cube(3^2);**

Which says "take 3, multiply it together 2 times, and use the result as the length of the sides of a cube. This is equivalent to:

> **cube(9);**

As is true in most computer code, * is used for multiply, / for divide, and + and - for addition and multiplication.

As we noted in Chapter 2, you can also always write your own function, too, which can use some of these other ones as well.

EXTRUSION

OpenSCAD has a few special functions that let you take 2D shapes into 3D, or combine objects. There are two ways to *extrude* objects, which means that you take a 2D shape and push it into the third direction, like toothpaste out of a tube, or homemade pasta out of a pasta machine. If you imagine the opening on the toothpaste tube being various shapes (more like homemade pasta out of a pasta machine), you can imagine how that works.

EXTRUDING LINEARLY

First, start with a 2D shape in the x-y plane, like **circle(...)** or **square(...)**, or combinations as complicated as you like. The OpenSCAD function:

> `linear_extrude(height, center, convexity, twist, slices, scale)`

Will take the 2D shape and make it **height** millimeters tall in the z-direction. To use linear_extrude, you first create a 2D shape and then extrude it. Only the **height** argument is required. If you just use one number in **linear_extrude** like this:

```
linear_extrude(5)
circle(15);
```

that's the height in the z direction. The example just given will create a circle 15mm in radius in the x-y plane and 5mm thick in the z direction. Technically, the height variable event isn't required. It defaults to 100mm.

What about those other variables? If **center = true,** the extrusion will go half in the +z direction and half in the -z direction. The variable **twist** allows you to twist the extrusion through that many degrees as it

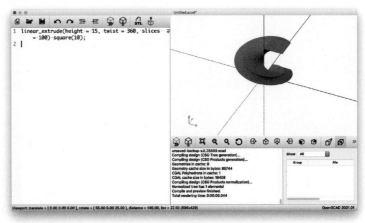

FIGURE 3-10: Linear extrude with a twist and more slices

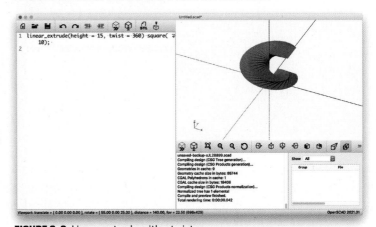

FIGURE 3-9: Linear extrude with a twist

goes up. If you set **twist = 360**, the extrusion will go through a full circle's worth of twist on its way up. Try this out with a square (Figure 3-9).

```
linear_extrude(height = 15, twist = 360) square(10);
```

The **slices** variable allows you to set how jagged or smooth the extrusion is in the upward direction, sort of like a third dimension for **$fn** that we discussed earlier in this chapter. This only matters if you are making your extrusion have a twist. A higher number is smoother (Figure 3-10).

```
linear_extrude(height = 15, twist = 360, slices = 100)
square(10);
```

The **scale** variable expands or contracts the top of your extrusion. The bottom of your shape will have the given dimensions, and the shape will

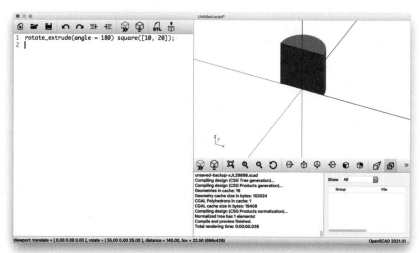

FIGURE 3-11: Half-cylinder

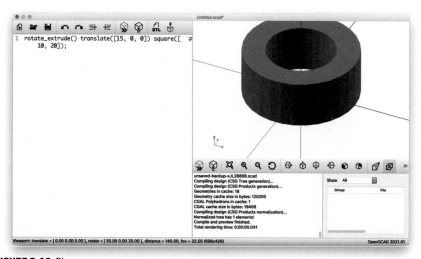

FIGURE 3-12: Ring

be smoothly scaled (bigger or smaller). For example, if **scale = 0.5**, the dimensions at the top of the extrusion would be half that of the bottom.

If you are going to be done with your model after extruding, you probably won't need to use the convexity variable. However, if you perform boolean operations (particularly **difference)** on the extruded object, you might see display glitches in preview mode. Parts of the model might look like they're inside out, or further away things appear in front of closer things. This won't affect your exported model, but it may make it difficult to see what's going on while previewing. Setting **convexity = 5** is usually sufficient to remove these glitches.

EXTRUDING WHILE ROTATING

If you want to make a shape that is a part of (or a whole) toroid (a shape like a donut with a hole) you can use the **rotate_ extrude(angle)** module. The extrude functions presume that you are starting with a 2D shape. First, you create a 2D shape like a square, which Open- SCAD makes in the x-y plane.

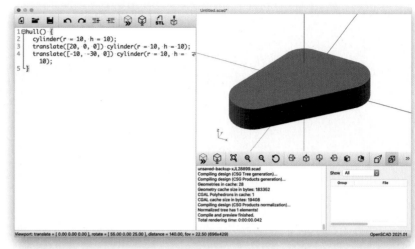

FIGURE 3-13: Triangular shape using hull()

The **rotate_extrude(..)** process works like this: the module first reorients your 2D shape from where you would, by default, make it in the x-y plane. It moves it to the x-z plane, then sweeps it around the z-axis in a circle (or part of one). A full circle would use **angle = 360**, and half would be **angle = 180**. The shape has to have entirely positive or negative x coordinates (not crossing the y-axis) for this to work.

For example, to get a (sort of jagged) half-cylinder (Figure 3-11) you would use this line of code:

```
rotate_extrude(angle = 180) square([10, 20]);
```

However, to get a ring with flat top and bottom (Figure 3-12) you would move the shape out from the z-axis first and then rotate it:

```
rotate_extrude(angle = 360) translate([15, 0, 0]) square([10, 20]);
```

You could actually leave out the **angle = 360**, since that is the default, and use this instead:

```
rotate_extrude() translate([15, 0, 0]) square([10, 20]);
```

It never hurts to use a default so you remember what you did later, of course.

CONVEX HULL

The **hull()** module combines multiple shapes into a continuous one, connecting across any gaps among them. It uses a mathematical construct called a convex hull to do this. All this means, really, is that it connects the shapes in ways that never sag inward into the shape. For example, you can create a set of points using cylinders, then fill in the space enclosed by them. Figure 3-13 shows the resulting triangular shape with rounded corners.

```
hull() {
  cylinder(r = 10, h = 10);
  translate([20, 0, 0]) cylinder(r = 10, h = 10);
  translate([-10, -30, 0]) cylinder(r = 10, h = 10);
}
```

This is particularly useful when combined with loops, allowing you to create things like a cube with its corners and edges rounded off (Figure 3-14).

```
size = 30;
r = 5;

translate([-size / 2, -size / 2, -size / 2]) {
  hull() {
    for(x = [r, size - r]) {
      for(y = [r, size - r]) {
        for(z = [r, size - r]) {
              translate([x, y, z]) {
                sphere(r);
              }
          }
        }
      }
    }
  }
}
```

Note that OpenSCAD doesn't have a way to speed up the hull operation for preview mode. Creating a hull requires calculating all of the geometry, so the preview may be slow. Once you calculate the hull's geometry, as long as you don't change anything inside it, OpenSCAD's cache should speed up subsequent preview and render operations.

MAKE A CASTLE

Let's try a project that will give you a workout both with math ideas in this chapter (basic 3D shapes, rotation, translation, and mirroring) along with working in OpenSCAD. We will walk you through the basics of a simple castle here. If you want to keep going, check out Chapter 13's medieval arches and windows, and see if you can figure out some fun ways to subtract some interesting windows or castle gates from one or more of the sides. This basic

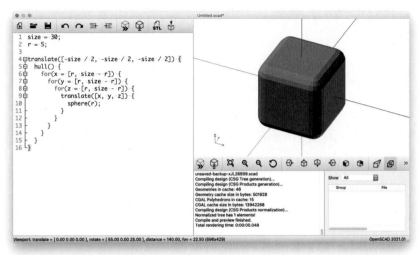

```
1  size = 30;
2  r = 5;
3
4  translate([-size / 2, -size / 2, -size / 2]) {
5    hull() {
6      for(x = [r, size - r]) {
7        for(y = [r, size - r]) {
8          for(z = [r, size - r]) {
9            translate([x, y, z]) {
10             sphere(r);
11           }
12         }
13       }
14     }
15   }
16 }
```

FIGURE 3-14: Rounded-off cube using hull()

design is in the file **castle.scad.** The final printed version is shown in Figure 3-19. You will notice that it has four sides that are the same.

THE WALLS

To start, we'll make just one of the four walls, and then duplicate it and move it around using **rotate(...)**, **translate(...)**, and the loop functionality we learned about in Chapter 2. Let's say our wall consists of a rectangle that is a variable **wall** long, **height** tall, and of thickness **thick**, all in millimeters. We are going to lay it out so that the ground the castle is on would be the x-y plane, where z = 0. Any height along a wall, then, will be in the z direction. That much would look like this:

```
wall = 100;
height = 40;
thick = 4;


cube([wall, thick, height]);
```

Now, any self-respecting castle needs crenellations along the top - parts of the wall that stick out that archers can hide behind. Let's make those out of cubes that are the wall thickness in all dimensions, and space them out by their width, too. (Yes, the archers would need a wall behind the crenellation to be able to walk, but you can add that sort of thing later!)

```
wall = 100;
height = 40;
```

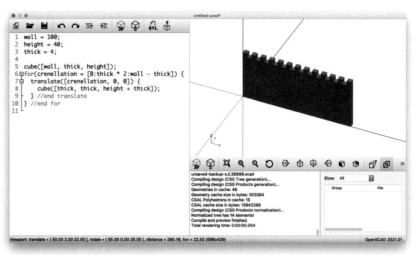

FIGURE 3-15: One wall and crenellations

```
thick = 4;

cube([wall, thick, height]);
for(crenellation = [0:thick * 2:wall - thick]) {
    translate([crenellation, 0, 0]) {
        cube([thick, thick, height + thick]);
    } //end translate
} //end for
```

This says: set my loop counter **crenellation** to zero; create a "cube" that is the same thickness as the wall, equally as long as it is thick, and taller than the main wall by the wall thickness. In other words, we have a cube that is **thick** on a side sitting on top of the main wall. Then, we do it again, moving **thick * 2** (the width of one of these cubes plus an open space the same size) down the wall, and so on. We do that until we reach the last place we can put one of these, which is when we have marched down the wall the distance **wall - thick**, which leaves just enough space for the last crenellation before the end of the wall.

Notice that we enclosed the code we wanted to execute multiple times in curly braces. If it's only one line of code, strictly speaking, you don't need those, but it is a good habit to get into. It is also considered good practice to indent lines of code that are inside a set of curly braces so it is easier to see what is enclosed, and to find the matching curly brace at the end. You may also note that we are combining the wall and the crenellations. Technically this is a **union()**, but OpenSCAD allows **union()** to just be implied (Figure 3-15).

THE TOWER

Now, what is a castle without some good guard towers at the corners? Let's make a tower for the corner of our wall. We'll start with a cylinder to have a nice round tower. Then we'll top it with a cone. OpenSCAD does

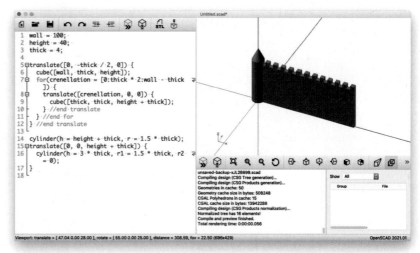

FIGURE 3-16: One wall plus tower, translated to be lined up with each other

not have a cone as such. What it does let you do, though, is define a "cylinder" that has different radii at the top and bottom, and call them **r1** and **r2** respectively. If **r2 = 0**, then the top is a point and we get a cone. The lines that produce a cylinder at the beginning of the wall with a cone on top are:

```
wall = 100;
height = 40;
thick = 4;

cylinder(h = height + thick, r = 1.5 * thick);
translate([0, 0, height + thick]) {
   cylinder(h = 3 * thick, r1 = 1.5 * thick, r2 = 0);
}
```

Cylinders are by default centered at x = 0 and y = 0, with their bottoms at z = 0. The tower overlaps the first chunk of the wall, but OpenSCAD doesn't mind that.

Next, we have to translate the wall plus its crenellations by half its thickness (**-thick/2**) so that its center line is aligned with the tower. Remember that if you use curly braces after **translate(...)**, everything inside the braces will be translated. This will line up the center line of our wall with our tower. (Figure 3-16). The part of the model that does this is:

```
wall = 100;
height = 40;
thick = 4;

translate([0, -thick / 2, 0]) {
```

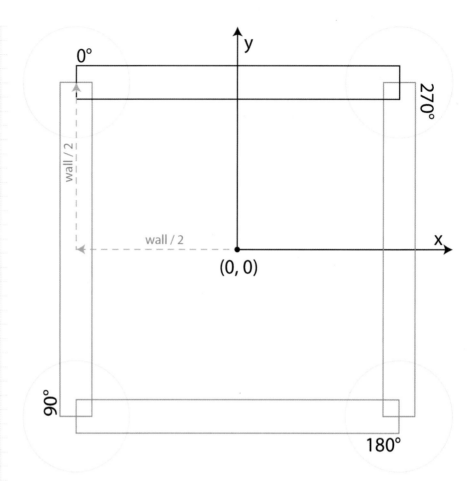

FIGURE 3-17: Castle setup in the x-y plane

```
cube([wall, thick, height]);
for(crenellation = [0:thick * 2:wall - thick]) {
  translate([crenellation, 0, 0]) {
    cube([thick, thick, height + thick]);
  } //end translate
} //end for
} //end translate

cylinder(h = height + thick, r = 1.5 * thick);
translate([0, 0, height + thick]) {
  cylinder(h = 3 * thick, r1 = 1.5 * thick, r2 = 0);
}
```

MAKING FOUR WALLS

We could just duplicate this code four times and do the appropriate translating and rotating for each one, but that would leave us with a lot of redundant code. It would also mean that if we want to make further changes, like adding windows, we would have to be sure to add four copies of that code, and keep them consistent with one another.

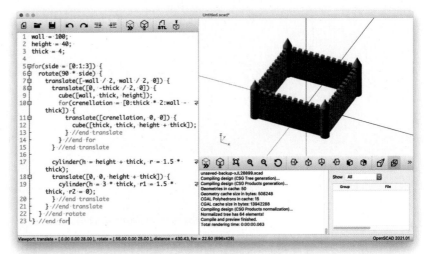

FIGURE 3-18: The whole castle (so far) in OpenSCAD

Instead, we will put the whole object inside yet another loop. To prepare our wall plus tower to rotate, first we will translate this entire collective object (wall plus crenellations plus tower) away from the center by **wall/2** in the x and y directions. For x, we'll need to move in the negative direction, so that the center of the wall is aligned with (0, 0). This original wall is at the top of Figure 3-17, and its tower is in the top-left. All that is left, then, is to make four copies of that wall, rotating around as we go (the other walls in Figure 3-17). OpenSCAD will rotate around the x = 0 and y = 0 point, so if we have translated our object first we will move around the center as shown.

Here is the entire model (downloadable as **castle.scad**). Figure 3-18 shows a completed print in OpenSCAD.

```
wall = 100;
height = 40;
thick = 4;

for(side = [0:1:3]) {
  rotate(90 * side) {
    translate([-wall / 2, wall / 2, 0]) {
      translate([0, -thick / 2, 0]) {
        cube([wall, thick, height]);
        for(crenellation = [0:thick * 2:wall - thick]) {
          translate([crenellation, 0, 0]) {
            cube([thick, thick, height + thick]);
          } //end translate
```

FIGURE 3-19: Photo of a print made from this model

```
      } //end for
    } //end translate

    cylinder(h = height + thick, r = 1.5 * thick);
    translate([0, 0, height + thick]) {
      cylinder(h = 3 * thick, r1 = 1.5 * thick, r2 =
0);
    } //end translate
  } //end translate
} //end rotate
} //end for
```

FANCIER CASTLES

Of course, there's no reason to stop here. You can add windows, fancier towers, a gate, and more. If you wanted to add holes for one doorway after the whole project was completed, you could do this:

difference() {
The model we just did
The shape of your doorway, translated/rotated as needed
}

This would subtract (create a hole) for the door. Or, if you wanted to create a window or door in all four walls, you could subtract it from the wall like this:

Define variables, set up wall rotation here

```
difference() {
  translate([0, -thick / 2, 0]) {
    cube([wall, thick, height]);
    for(crenellation = [0:thick * 2:wall - thick]) {
      translate([crenellation, 0, 0]) {
        cube([thick, thick, height + thick]);
      } //end translate
    } //end for
  } //end translate
```

The shape of your window, translated/rotated as needed

```
}
```

In Chapter 13, we will explore a little about how medieval architects used geometrical concepts and simple tools, and we will walk you through a renovation of this simple castle.

FIGURE 3-20: Raising a piece above the workplane in Tinkercad

TINKERCAD

You can use Tinkercad to explore translation and rotation, too. To translate something in the x and y directions, you just drag it. To raise something away from the workplane, in the z direction, click on it, then click the black triangle that appears above it (Figure 3-20) and drag the piece upward.

To rotate something in Tinkercad, first drag out the ruler (Figure 3-21).

Select the object again and you will see curved arrows near the corners (Figure 3-22). If you click on one of those arrows, you can rotate the piece around its axis by dragging along the circle, or you can type in the number of degrees you want to turn (Figure 3-23). Select a different view (by clicking the view cube at the upper left) and do the same thing in the new view to rotate about a different axis.

Tinkercad's code blocks also allow you to rotate and translate a part. The rotation or translation (which Tinkercad calls "moving") modifiers are placed *after* the definition of the object. Figure 3-24 is an example of an extruded star, which starts out laying on the workplane (the x-y plane for 3D printing). It is then rotated 45 degrees around the y axis and moved up 20 mm in z. Unlike OpenSCAD, the object rotates about its center, regardless of where it is. As we mentioned in the last chapter, this capability is being developed and

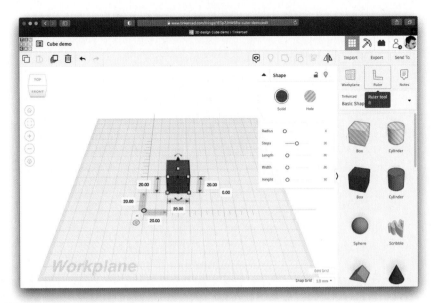

FIGURE 3-21: Tinkercad's ruler

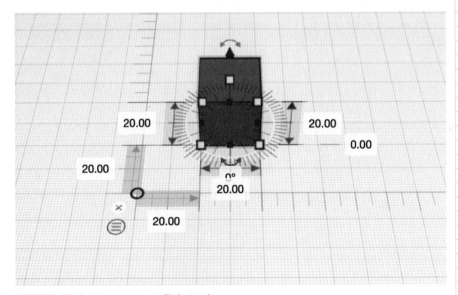

FIGURE 3-22: Rotation arrows in Tinkercad

you can play with their tutorials to see if there are other visualizations that you might find helpful.

SUMMARY AND LEARNING MORE

In this chapter, we went deeper into OpenSCAD and also showed how you can directly use it to simulate concepts like translation and rotation. We reviewed an

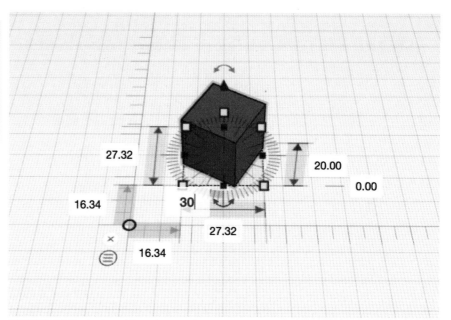

FIGURE 3-23: Typing in numbers for rotation

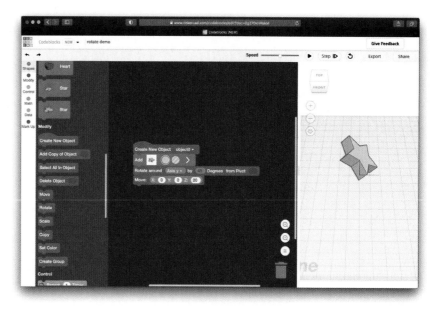

FIGURE 3-24: Rotating and translating in Tinkercad code blocks.

existing model and how to alter it, and created a castle model to demonstrate how to build up a complicated model from just a few basic structures.

As we have said frequently during the chapter, the best ways to learn more about OpenSCAD are to read its manual at openscad.org, but also to play with it. Try things out, see how the models can break, and then try to fix them. It's far and away the best way to learn.
In addition to simulating geometrical concepts, we also dove into polygons and polyhedra. There are many open resources to explore (starting with the Wikipedia articles with those titles).

In the next chapter, we will move to a more analog device: the compass and ruler, to see some alternate ways of constructing geometry.

CHAPTER 4
CONSTRUCTIONS

In the last two chapters, we learned how to use OpenSCAD to simulate various geometrical shapes and transformations. We also saw how those abstract models could become a 3D print. In this chapter, we'll learn a very old and more concrete way of simulating and creating geometrical shapes, the use of a tool to draw straight lines (like a ruler), and the use of a drawing compass.

We'll get into more detail on the properties of what we are constructing—angles, perpendicular lines, triangles—in future chapters. In this chapter, we want to get across the power of these simple tools, and how elegantly you can use them to construct accurate drawings.

CONSTRUCTIONS

We've mentioned the ancient Greek mathematicians a lot, since they developed the basics of geometry. Some of it was done as an intellectual exercise. However, much of the development of geometry was an early attempt to solve practical problems like navigation, specifically finding latitude (see Chapter 7). One of the best-known mathematicians of the group was Euclid. He lived about 2400 years ago, and his geometry book *Elements* has more or less been continuously in print ever since. If you search for "Euclid Elements" you can find various free versions and translations, or you can buy more modern versions.

Obviously, the ancient Greeks were very limited in the tools they had at hand. They worked with a ruler (more formally called a "straightedge"), a compass, and some way to mark a surface. We have the advantage of pencil and paper, but otherwise we can try to get some of their insights the same way they did.

We will give you a few examples, and we hope you play around with them and find other relationships. If you really enjoy this but would prefer something more like a video game, you can try playing the game *Euclidea*. It is available on the web (**https://www.euclidea.xyz**) and as a phone app. It simulates the use of a compass and straightedge, and combines some of these tools so you can skip some of the steps. However, you will probably enjoy the game more if you learn a little geometry first, since Euclidea doesn't give many hints!

Now, we will walk through the steps of doing three classic constructions. We'll show one particular type of compass because it is a little easier to see what is going on with this type, and it is a bit easier to swing around as you will see. We'll refer to the *needle point* of the compass as the point that does not have a pencil on it,

3D Printable Models Used in this Chapter

See Chapter 2 for directions on where and how to download these models.

reuleaux-n-gon.scad
Creates objects called Reuleaux polygons (explained in this chapter)

Other supplies for this chapter
- Compass (the kind that draws circles)
- Ruler (straightedge)
- Pencil and paper
- Alternatively, string

How to Use a Compass

A compass is an instrument for making circles, but it can also be used to do constructions, which is what mathematicians call it when you draw something precisely, often to prove that a particular statement is true.

There are several compass designs, but most hold a pencil in a way that makes it easy for you to make a circle. We will show you three types of compass you might encounter.

The most common kind somehow attaches or incorporates a pencil on one side. You hold the handle on top, and pivot the pencil around the center. This type (Fig. 4-1) is the easiest to use precisely, and it is simpler to be able to pivot from one circle to another. It is also the easiest to see what you are doing, so we will use this type in future illustrations.

FIGURE 4-1: A standard compass

More recently, there are plastic "safety compasses" of varying design (Fig 4-2). We show two of them here. You hold down the center of a flat piece of plastic, letting the arm turn freely. Smaller circles can be made by using the cutouts in the middle of the arm.

In a pinch, you can use a piece of string with a loop in it: put the pencil against the loop, hold down the other end of the loop as the center of the circle, and draw. However, we are going to try to do some things that require a bit of precision, and it will be a lot easier if you can get an actual compass.

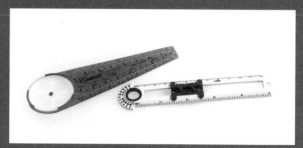

FIGURE 4-2: Safety compasses

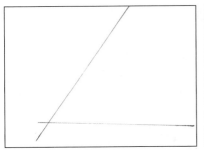

FIGURE 4-3: First, start by drawing two lines with the straightedge to create an angle

FIGURE 4-4: Now, take your compass and put the needle point on the vertex of the angle.

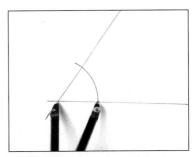

FIGURE 4-5: Swing the pencil point around the needle point to draw an arc centered on the vertex

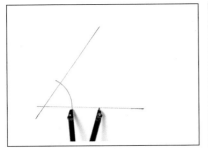

FIGURE 4-6: After you have drawn your arc, move your compass so that the needle point is on the intersection of the arc and one side of the angle

FIGURE 4-7: Now draw another arc farther from the vertex.

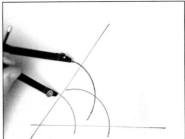

FIGURE 4-8: Figure 4-8: Repeat this move from the other side of the angle.

and the *pencil point* as the one that does. For the compass with a rotating central part, obviously the middle of that is the equivalent of the needle point. We'll use a protractor (normally used to measure angles, as we will see in Chapter 5) as a ruler here since it is easy to see through it. However, any ruler will do.

BISECTING AN ANGLE

This construction cuts an angle in half, also known as *bisecting* an angle. We will talk more about angles and measuring them with a protractor in Chapter 5, but now just focus on how we can pretty accurately cut one in half.

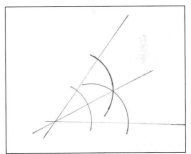

FIGURE 4-9: Now draw a line with a ruler from the vertex of the angle to the intersection of the two arcs you just drew. That line will cut the angle in half.

Be sure to keep the two points of the compass the same distance apart for the whole drawing. The compass in our picture has a locking mechanism, but if yours does not you might just need to be careful not to spread it out or compress it. Figures 4-3 through 4-9 detail the process. The radius of the arc doesn't matter, although you should have it intersect the sides of the angle fairly close to the vertex so you have room for the later steps.

Why this works: Circles are used in constructions a lot because, since a circle's radius is constant (as we will explore in depth in Chapter 7), they are a handy way of measuring that two distances are the same. When you drew the first arc, you marked points the same distance from the vertex on both sides of the angle. The vertex is an equal distance from each of the two points where the arc crosses the sides of the angle.

Because the bisecting line runs down the middle, every point on it is also equidistant from those two points. By drawing two equal-radius arcs, one centered on each of those two points, we can find another point on the bisecting line where they cross. Connecting this crossing point to the vertex gives us the bisecting line.

PERPENDICULAR BISECTOR

Here, we are finding a line that cuts a given line in half, at a right angle to it. Figures 4-10 through 4-13 walk through the details of the process.

Why this works: As in the previous example, a circle's radius is constant, and every point on the bisecting line is an equal distance from the two endpoints. Since a slanted line would have to pass closer to one endpoint than the other, this bisecting line has to be perpendicular. When you draw equal-radius arcs from either end of the line, the places where they cross have to be equidistant from those points. When you connect the two points where the arcs cross, you get a perpendicular line of points that are all equidistant from those two centers.

FIGURE 4-10: Adjust the width of the compass by putting the needle point at one end of the line you want to bisect, and the pencil point at the other.

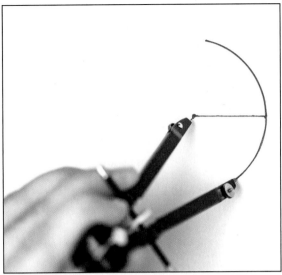

FIGURE 4-11: Draw an arc with the pencil point around the needle point a bit beyond where you estimate the midpoint of the line is — say about two-thirds of a circle.

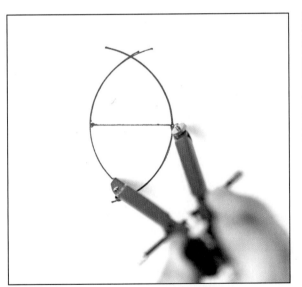

FIGURE 4-12: Leave the compass points the same width apart, but now put the needle point on the other end of the line you are bisecting. Again draw an arc until it intersects the one you just created.

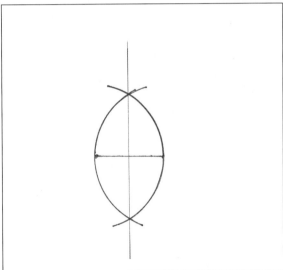

FIGURE 4-13: Now use a ruler to draw a line that intersects the two places where the arcs you drew in the last two steps cross each other. That line is the perpendicular bisector.

FIGURE 4-14: First, set your compass points a convenient distance apart.

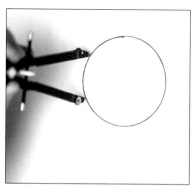

FIGURE 4-15: This distance will be the length of your triangle's sides, and you will leave them that way for the rest of the process. Start by drawing a circle.

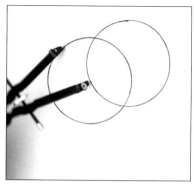

FIGURE 4-16: Put the needle point at any point of the circle you just drew and draw another, intersecting circle of the same radius there.

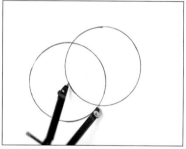

FIGURE 4-17: Now put the needle point at one of the places where these circles intersect. Draw a third intersecting circle.

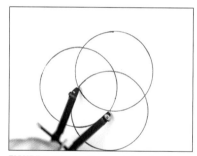

FIGURE 4-18: You'll see a triangle-like shape in the middle. In the next section, we'll discover it is called a Reuleaux triangle.

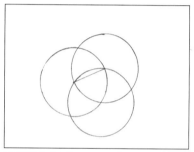

FIGURE 4-19: Use your ruler to connect two vertices of this equilateral triangle.

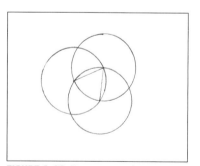

FIGURE 4-20: Next, use a ruler to connect two more vertices.

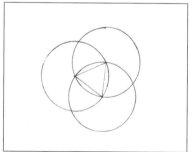

FIGURE 4-21: Connect the final two vertices with a ruler.

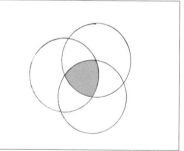

FIGURE 4-22: A Reuleaux triangle (grey) as constructed

EQUILATERAL TRIANGLE

This construction allows you to create an equilateral triangle, which has all its sides (and, therefore, all of its angles) the same. We'll learn more about them in Chapter 5. Figures 4-14 through 4-21 show all the steps.

You will now have an equilateral triangle. There are also three more possible equilateral triangles of the same size in the overlapping circles you have drawn. Where are they? You might even observe that there is another, larger equilateral triangle you could have drawn in this last step. Where is it, and what is the ratio between its size and the triangle in Figure 4-21? See the "Answers" section at the end of the chapter to check yourself.

Why this works: As in the previous constructions in this chapter, a circle's radius is constant. Because an equilateral triangle's sides are all equal, each vertex is equidistant from the other two. We start the process similar to the previous example, where we were creating a perpendicular bisector. The points where the first two circles cross gives us two points that are equidistant from their respective centers, and connecting the centers to each other and to one of those points gives us our triangle.

If we had been careful to mark those centers, or if we had drawn one line of the triangle first and used points on that line as the centers, we wouldn't even need to draw the third circle. Since those first two points are the same distance from the third point, though, drawing the third circle, which passes through both of their centers, will make it a little more clear where those centers were by intersecting with the other circle at each center.

OTHER CONSTRUCTIONS TO TRY

You can find many other constructions if you want to try playing with them; the chapter summary suggests more places to go. You might think about how to construct a hexagon next, and search on it to see if you were right! How might you use constructions if you were an architect laying out a house? Or if you were living in medieval times and wanted to measure out a building with a rope as your compass and a taut string as your straightedge, how would you do that? Look ahead to Chapter 13 if you want to play with some of these ideas.

REULEAUX TRIANGLES

If someone asked you what shape had the same diameter everywhere you would probably answer that it was a circle, as we will explore in Chapter 7. However, there are other constant-diameter polygons. Note that we say constant *diameter*, not radius, since it is possible to have constant-diameter objects that are not symmetrical around a center point. One of the best-known ones is the Reuleaux triangle (pronounced roo-low). A Reuleaux triangle is not a true triangle, in that its angles do not add up to 180° like other triangles do, as we'll see in Chapter 5. In fact, its internal angles are all 120°, and so they add up to 360°.

It's named after the German engineer Franz Reuleaux, who was active in the mid-to-late 1800s. He invented a lot of what we now call *kinematics*, which is figuring out how a mechanism would work. His diagram-filled 1876 book, *The Kinematics of Machinery*, is available at **https://en.wikisource.org/wiki/The_Kinematics_of_Machinery**, if you want to play with some of his other constructs.

We can get a Reuleaux triangle pretty easily as a byproduct of our equilateral triangle construction. To make a Reuleaux triangle out of paper, start the equilateral triangle construction in this chapter and go as far as drawing the three circles. You will see something that looks like a rounded triangle in the middle (Figure 4-22). That's the Reuleaux triangle.

Use your compass to sketch out one of these on stiff cardboard and then cut it out for the explorations to follow. If you have a 3D printer, you can use the **reuleaux-n-gon.scad** model.

3D PRINTABLE MODEL

The 3D printable model has a few variables you can play with. It creates a Reuleaux polygon (Reuleaux polygons exist for all odd numbers of sides) and an enclosure. We'll talk about that enclosure in a minute. Meanwhile, here are the variables and their defaults.

> **width** = 50;
> > Width of the polygon, mm
>
> **sides** = 3;
> > Number of sides (has to be odd)
>
> **thick** = 10;
> > How thick the polygon is when printed, in mm
>
> **wall** = 1;
> > Wall thickness of the two pieces, in mm
>
> **base** = 0.6;
> > Thickness of the solid base on each piece, in mm

We'd suggest leaving the defaults alone, other than changing the number of sides. If you want to scale the model, however, do it by changing the **width** variable, rather than by scaling in a slicer. Otherwise the side walls might get too thin or be too thick for the rotation of one model in another (which we'll talk about next) to work correctly.

FIGURE 4-23: A Reuleaux triangle inside its square enclosure.

ROLLING REULEAUX

We know that round wheels will roll, but it turns out that's true of any constant-width shape like a Reuleaux triangle. The width is constant, but the distance to the center from all points on the perimeter is not equal. This means that you could in principle make a wheel that is shaped like a Reuleaux triangle, but it would need a complicated mechanism to work since the center of rotation has to move.

People have made bicycles with Reuleaux triangle wheels (search online to see examples, often spotted riding around places like the Burning Man event) but they are pretty complex. Rather than rotating around a fixed hub, the centers of the wheels are allowed to move up and down in a way that requires a complex mechanical linkage. If you create a Reuleaux triangle, either by 3D printing or cutting one out of cardboard, you can see that it rolls easily along a flat surface, but is not rotating around any one constant point.

What happens if we confine a Reuleaux triangle in an appropriately-sized square (Figure 4-23)? It will turn freely. Actually, if you were to make a Reuleaux triangle drill bit, it could be used to drill out a square hole with slightly rounded corners. 3D print a Reuleaux triangle and its square enclosure, and rotate the triangle. You'll see it will go almost completely into the corners of the square.

FIGURE 4-24: A Reuleaux 5-gon.　　　　　**FIGURE 4-25**: A Reuleaux 7-gon.

Similar features show up more often than you might think, some of them long before Reuleaux analyzed them. In Chapter 13, we will see how Gothic arches in medieval cathedrals relate to Reuleaux triangles, and construct 3D printable outlines of cathedral window designs by using related constructions.

OTHER REULEAUX POLYGONS

The 3D printed model provides both a Reuleaux shape of varying numbers of sides (change the variable "sides" to an odd number) and an enclosure which is approximately the shape they make when they are turned in place. Print out several and turn the Reuleaux shape around to see! Here you can see what a 5-gon (Figure 4-24), and a 7-gon (Figure 4-25) look like.

Constant-radius shapes are pretty handy because they can fit into machinery well, like coin machines. In fact, the British 20p and 50p coins are Reuleaux 7-gons (Figure 4-26). Using different shapes makes it easier to distinguish one coin from another. It is true for all odd-sided Reuleaux polygons that they make a hole with one more side. Even-numbered Reuleaux polygons don't exist, since you need to have an arc of a circle opposite each vertex to have the constant-distance construction work out

In Chapter 13, we'll also introduce you to the Reuleaux tetrahedron and its cousins the Meissner tetrahedrons. These are 3D versions with their own strange properties.

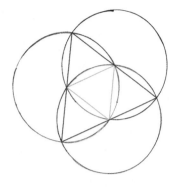

FIGURE 4-26: British coins that are Reuleaux 7-gons

FIGURE 4-27: The other equilateral triangles that can be constructed

SUMMARY AND LEARNING MORE

If you enjoyed the constructions, we recommend searching for a copy of Euclid's *Elements* and seeing how much they knew 2500 years ago. You also might look at the Wikipedia entries for "straightedge and compass construction." If learning from a book isn't your thing, there are now many websites and games that simulate the process of doing geometric constructions. We already mentioned Euclidea (**www.euclidea.xyz**) which is more of a game than a course. There are also more traditional explanatory sites, like GeoGebra (**www.geogebra.org**).

There are many places you can go from here. In Chapter 13 we will see how the constructions of circles and Reuleaux triangles can be used to draw elements of Gothic (and later) architecture. Meanwhile, the next three chapters will introduce you to two-dimensional shapes, starting with triangles in Chapter 5, and we'll fill in some of the properties of the shapes you've just been constructing.

ANSWERS
EQUILATERAL TRIANGLE CONSTRUCTION

In Figure 4-21, we showed the construction of a central equilateral triangle, shown in blue in Figure 4-27. There are also three more "outer" triangles, spaced evenly around the first one. These outer triangles have one blue side and two red ones.

All four equilateral triangles together form a larger (red) equilateral triangle. (Figure 4-27). Since all four of these smaller triangles are equal, we know that the length of each side must be exactly twice that of the smaller triangles. So the ratio of the size of the "inner" (blue) and "outer" (red) triangle is 1:2.

5

CHAPTER 5
THE TRIANGLE BESTIARY

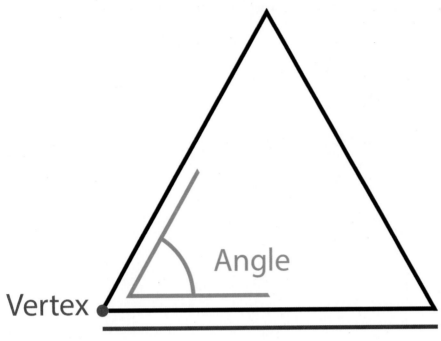

FIGURE 5-1: The anatomy of a triangle

In medieval times, people wrote books called "bestiaries" about weird and wonderful beasts and what lessons one could learn from them. Triangles are deceptively simple and show up where you least expect them. Every shape that you can make with straight sides can be broken into triangles. They make up the surface of a 3D print, the structure of a bridge, and the roof of a house. Let's look for these beasts in the wild, see what makes them operate the way they do, and analyze their habitat.

WHAT IS A TRIANGLE?

A *triangle* (Figure 5-1) is the simplest way you can make a shape with straight lines that has a distinct inside and outside. It is made of three lines. (Usually those lines are straight, but we saw an example of when they aren't with the Reuleaux triangle in Chapter 4.) Mathematicians call the points where the lines meet the *vertices* (just one is called a *vertex*).

ANGLES OF A TRIANGLE

Back in Chapter 3, we learned about angles and how they are defined. Triangles have a special relationship with some angles. Triangles with one 90° angle are called *right triangles*, and a little later in this chapter we will learn more about special ratios of their angles and sides. Two times 90° is

3D Printable Models Used in this Chapter

See Chapter 2 for directions on where and how to download these models.

TriangleAngles.scad
Demonstrates that the angles of a triangle sum to 180 degrees

ExtrudedTriangle.scad
Just two lines long, this model draws a triangle from any three points and extrudes it to the desired thickness.

TriangleSolver.scad
Creates three similar triangles (scaled). They can be specified using one of several rules, described in the text.

TriangleArea.scad (TWO SETS needed)
This model consists of a triangle broken into several pieces. It has an option to print out one or two triangle sets. Set the variable to TRUE to make two sets, or FALSE to make one.

TriangleBox.scad
This model creates a rectangular open box to support the TriangleArea triangles.

TriangleAreaBox.scad
This model creates a triangular open box that holds together just one TriangleArea triangle.

Other supplies for this chapter

- Cardboard
- Paper clips
- Protractor
- 0.2mm stretch cord
- Drinking straws
- A few pieces of construction or similar heavy paper
- 3 rulers (we used 8-inch ones, but length isn't important)

FIGURE 5-2: The triangle assembled

180°, and a straight line is a 180° angle. If you kept going all the way around to bring your hand back to where you started, you would have turned 360°, or a full circle.

If you think about it, as one angle of a triangle gets narrower, the triangle will be pinched together at that vertex, but the other two angles will need to get correspondingly bigger so that it stays a triangle. We can either use the 3D printed model **TriangleAngles.scad** or just a triangle cut out of paper to demonstrate that, as it turns out, the angles inside a triangle always add up to 180 degrees. Let's use a very old trick that dates back to Euclid (who lived around 2400 years ago in Greece).

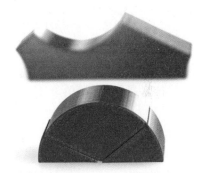

FIGURE 5-3: Figure 5-3: The triangle's vertices rearranged to show that the angles sum up to a straight line.

PROVE TRIANGLE ANGLES ADD TO 180 DEGREES

Make a triangle cut into four pieces, either by 3D printing the TriangleAngles.scad model or using paper. To make a paper version, first cut out any triangle you like. Then, using a compass, cut off each of the triangle's vertices (as we show with the 3D printed triangle in (Figure 5-2)). You'll need to keep track of which corner of each piece was a vertex of the original triangle, and making a curved cut as we have done here will make that easier.

Now remove these three vertex pieces from the triangle and put the three vertices together, so they make a straight line (and thus add up to 180 degrees). You can see this in Figure 5-3.

The longest side will have the biggest angle directly across from it, and the smallest side will have the smallest angle across from it. Of course, that means that if two angles are equal, the sides opposite

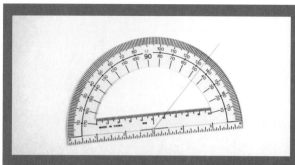

FIGURE 5-4: Measuring an angle with a protractor. (Base of the angle is parallel to the line across the bottom of the protractor.)

How to use a protractor

A protractor is a device for measuring angles. You can print one out (search online for "download protractor") or you can buy one. If you buy one, a clear plastic one is handy because it is easier to see what you are doing.

If you are measuring the angles of a triangle with a protractor, first put the vertex of the angle you are measuring on the crosshairs at the bottom of the protractor. Line up the bottom of the angle with the line on the bottom of the protractor(Figure 5-4).

Then you can read off the angle from the scale around the edge. Either read up from the bottom if the angle is less than 90° (as in this case, where the angle is about 47°) or the outer scale if it is more than 90°. Note that how carefully we can measure comes down to how good our tools are. The width of the lines making up your triangle, how accurately your protractor is printed, and how good you are at estimating will all come into play. A plastic protractor like the one shown is probably good to plus or minus half a degree.

them will be the same length. Print, or cut out, several variations of this model (changing the angles each time) to get some intuition about why this is true. You'll see that if some of the angles get wider, the others have to become correspondingly smaller so that the sum of the angles will remain the same.

For right triangles (with, by definition, one 90° angle) this means that the other two angles have to add up to 180° - 90°, or 90°. Since each of these other angles has to be less than 90°, the side that does not touch the 90° angle will be the longest. This longest side of a right triangle (the one opposite the right angle) is called the *hypotenuse*. The longest side (opposite the biggest angles) doesn't have a special name in other triangles.

SPECIFYING TRIANGLES

How many angles and/or sides of a triangle do you need to know to pin down all its dimensions and angles? Try to figure it out before you read the answer in the next section.

CONGRUENT TRIANGLES

Congruent triangles are the same size and have the same angles as each other (although they can be oriented differently). If we 3D print a triangle or cut one out of a piece of paper, it is pretty obvious that if we take this plastic or paper triangle and flip it over (mirror it), rotate it around its center, or move it (which mathematicians would call "translating"), the triangle does not change. If we were to do any of those things and trace around the triangle in the starting position and then trace around the triangle after we had rotated, translated, and/or mirrored it, each of those traced triangles would be *congruent* to the one we started with.

You can't stretch the triangle and have it stay congruent, though. The congruent triangles are all ones that could be drawn around the same physical triangle, which means all the sides and angles of the triangle have to stay constant.

But you don't need to know all three sides and all three angles to say that two triangles are congruent. If you know two angles of a triangle, you know all three (because they have to add up to 180°, as we learned in the last section). When you've set the three angles of a triangle, all of the triangles you can make with those angles will be *similar* triangles. They will be the same shape, but because none of the lengths have been provided, they will not necessarily be the same size (and therefore, not congruent).

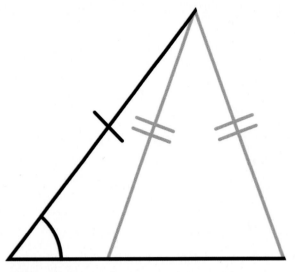

FIGURE 5-5: Showing two different triangles can be constructed for a given side-side-angle.

However, if you know the lengths of two sides and the angle between them (the "included angle"), that's enough to specify the whole triangle. This is known as the "side-angle-side" (or SAS) method of seeing if two triangles are congruent. Knowing all three sides will do it too (SSS) since that completely determines the possible angles of the triangle. Knowing two angles and their included side (ASA) is another option, as well as one side, one angle adjacent to it, and the opposite angle (SAA).

However, some combinations don't work. Just having all the angles ("angle-angle-angle", or AAA) determines the triangle's shape, but not its size. SSA (two sides and the angle that is NOT between them) does not always define a triangle uniquely, as we can see in the example in Figure 5-5. With the angle and the side on the left (the one with one hash mark), and a second side the length of the two equal-length lines (shown with two hash marks), we can construct two different triangles that fit the criteria. Obviously, they are different triangles, even though they both have an angle in common and two sides of the same length.

THE TRIANGLE FAMILY TREE

You can prove the relationships in the previous section by getting three rulers and arranging them to make triangles. For example, if the triangle is a *right* triangle, you already know there is one 90° angle. If you set one of the other two angles and the included side (angle side angle, as discussed in the last

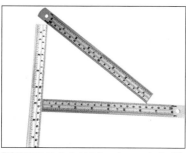

FIGURE 5-6: A right triangle created with three rulers.

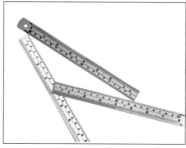

FIGURE 5-7: The triangle in Figure 5-6, with the right angle widened out

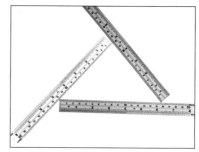

FIGURE 5-8: Figure 5-8: Acute triangles have all angles less than 90°.

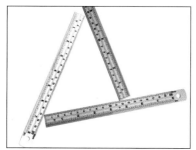

FIGURE 5-9: Figure 5-9: Scalene triangles have all three sides different lengths

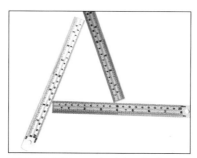

FIGURE 5-10: Isosceles triangles have just two sides the same length

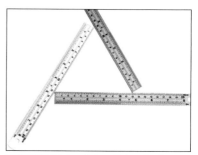

FIGURE 5-11: Equilateral triangles have all three sides the same length

section), you can only lay down the third ruler one way to make a triangle (Figure 5-6).

If you were to widen out the right angle, you would see that the two rulers making that angle would splay apart, and the opposite side would be determined (and longer than the hypotenuse was when it was a right triangle (Figure 5-7). This is called an *obtuse triangle*, with one angle more than 90°.

If we were to narrow down the angle between the rulers, we would get an *acute* triangle. Acute triangles have all angles less than 90° (Figure 5-8).

These three categories were determined by the triangle's largest angle. We can also categorize triangles by the relative lengths of their side as *scalene* (all three sides different length, Figure 5-9); *isosceles* (two sides of the same length, Figure 5-10); and an *equilateral* triangle, with all three sides the same length (Figure 5-11).

Earlier in this chapter, we saw that the angles of a triangle add up to 180°. That means that the remaining two angles of a right triangle would add up to 180° - 90° = 90°. This tells us that right triangles can't be acute or obtuse. The definitions of acute and obtuse mean that a triangle can't be both, and so we have three categories that don't overlap.

In the same way, the definitions of equilateral, isosceles, and scalene do not overlap. Some people like to say that isosceles triangles have *at least* two sides the same, and consider equilateral triangles a special case of isosceles, but we think that is confusing and will not use the words that way.

If two sides of a triangle are the same length as each other, the two angles opposite those sides are the same, too. We could also think of equilateral triangles as having all three angles the same, isosceles having two angles the same, and scalene as having all angles different.

What about the mixes of these categories? Some of them are possible, some are not. Try to fill out this table by using the three rulers or reasoning with a pencil and paper. To get you started, we just saw that a right triangle has one angle fixed at 90°, but the other two angles can be any two that add up to 90°. That means that the three sides can be different lengths from each other, too. So we would fill in a "YES" here to show that it is possible to have a right triangle that is also scalene. Fill in the rest of these and check your work against the filled-out table at the end of the chapter.

Triangle Types	Acute	Obtuse	Right
Equilateral			
Isosceles			
Scalene			YES

MAKING TRIANGLES WITH OPENSCAD

The OpenSCAD model **ExtrudedTriangle.scad** is just two lines long:

```
thickness = 10;
linear_extrude(thickness) polygon([[0,0], [10,20],
[20,10]]);
```

It creates a 3D extrusion of an example of a scalene triangle defined by the three points listed. In this example, it would draw a triangle with vertices at the points [0, 0], [10, 20], and [20, 10]. You can make any triangle you like by entering its vertices. The **thickness** variable determines how many millimeters the triangle is extruded. We have found 10 mm (a little less than half an inch thick) is a good value.

Drawing and Labeling Triangles

It can be convenient to note whether the sides and angles of a triangle are equal to each other. Often in geometry, we don't care what the actual measurement is; we just want to know which sides or angles are equal. Different authors use different conventions. A very common one that we will use is to show one, two, or three little crossbars on the side of a triangle. If two sides are the same, they will have the same number of bars. The angle across from that side gets the same number of little arcs. For instance, an isosceles triangle is shown in Figure 5-12, a scalene one in Figure 5-13, and a right triangle in Figure 5-14.

Sometimes we may label an angle or a side with letters or their numerical values, too. You may also see some books that draw a triangle like the one in Figure 5-14.

Now let's talk about angle ABC. Some books would call this angle B. The bottom line is that you'll need to check what the book or website you are using is doing. We use the convention of little hatch marks, as in Figure 5-14, except for a few places where it is clearer to do it the Figure 5-15 way.

FIGURE 5-12: An isosceles triangle labeling corresponding angles and sides (two equal)

FIGURE 5-13: A scalene triangle labeling corresponding angles and sides (all unequal)

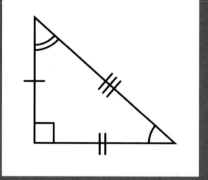

FIGURE 5-14: This scalene right triangle identifies the right angle as with a little square.

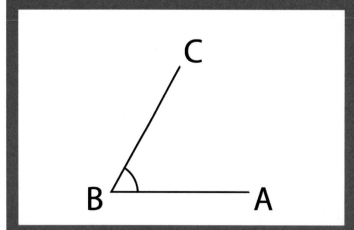

FIGURE 5-15: How to show a labeled angle

It can be a little annoying to work backward from the desired dimensions of a triangle to the coordinates of its vertices, so let's get a bit more sophisticated with the **TriangleSolver.scad** model. This model allows you to specify a triangle by any of the rules we discussed in the previous section (SAS, SSS, ASA, and SAA). Angles are in degrees in OpenSCAD, and sides are in millimeters.

The program makes a set of three triangles with the same angles as each other, but scaled up or down. These are called *similar triangles*, and we will talk about them in the next section. To use **TriangleSolver.scad**, you need to change two lines of the model to make your desired triangles. (If you skipped Chapters 2 and 3, which explore OpenSCAD extensively, you might want to go back and review those before going further here.) These two lines (which have other lines intervening in the model) are:

```
trianglebase = 30;
```
and
```
thicken() scaleset(trianglebase, [.5, 1, 2]) sas(30, 45,
30);
```

The parameter **trianglebase** tells the program how to avoid overlapping the three similar triangles. This parameter should be equal to the base of the triangle, which the program assumes is the first side given (or, for SAA or ASA, the only side.) The three numbers following are the scaling relative to the given length of a side. In this example, **[.5, 1, 2]** says that we will create a triangle that is half the dimensions of the original triangle, one that is the original dimensions, and one that is twice the size. The program allows you to list as many scaling factors as you like (but they will be lined up so you need to think about what will fit on your printer).

Finally, you can change how you specify the triangle by changing **sas(30, 45, 30)** (which would give you a triangle with two sides 30 mm long separated by 45°) to one of the other options - **sss(...), asa(...)** or **saa(...)**).

SIMILAR TRIANGLES

If two triangles have two angles the same, the third angle must be the same as well. The triangles might be different sizes, but regardless of the size, the ratios of the sides will be the same. These are called *similar* triangles. TriangleSolver.scad creates three similar triangles. You need to specify one triangle and then the set of scaling factors is applied to create the desired number of triangles, as described in the last section. Another quick way to

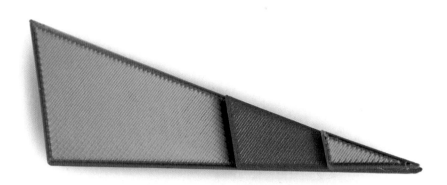

FIGURE 5-16: Similar triangles lined up to show their angles are the same.

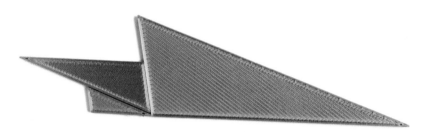

FIGURE 5-17: The three similar triangles in Figure 5-16 rearranged to show that taking one angle from each triangle still adds up to 180°.

prove that the three angles are the same in all three triangles is to overlay them, with the relevant angle lined up.

Note that similar triangles have the same set of angles, but are not (necessarily) congruent because they may be different sizes, as we see here. If the sizes of the triangles were the same, they would indeed be congruent, so congruent triangles are a special case of similar triangles. Here we can see these three triangles all have at least one angle in common (Figure 5-16).

Mixing the three scaled triangles still allows you to create a straight line out of the angles, just as we saw in the exercise where we lined up the three angles of one triangle earlier in the chapter. In other words, the angles add up to 180° even if you use the angles from the same triangle scaled up or down.

To test this out for yourself, take a set of similar triangles created by **TriangleSolver.scad** (or that you create with paper and a protractor) and show that even though the scales of the triangles are different, all three angles are the same (Figure 5-17). We took three similar triangles (which each have the same three angles as each other), and we turned each triangle so that we

FIGURE 5-18: Making a right triangle from a square piece of paper

used one different angle from each one. So, it is equivalent to cutting off the three angles of one triangle and aligning them, as we did in Figure 5-2 and 5-3.

AREA OF A TRIANGLE

We've been talking a lot about a right triangle. Try folding a piece of square paper along its diagonal to make a right triangle (Figure 5-18).

The area of the square piece of paper is just the lengths of the bottom (or top) and a side multiplied together. Since you folded one half of the paper over on itself, the area of this triangle is just one-half of the area of the square piece of paper you started with.

But what about all the other triangles? Let's demonstrate that the same general procedure applies more broadly no matter what the shape of the triangle.

TRIANGLE AREA MODELS

We have created several models that you can use together to visualize how several different triangles, all of which have the same base and height, will occupy half of a rectangle with the dimensions base * height. We do that by slicing up the triangles so that it is possible to arrange two of them into the rectangle. If two of them fit exactly into the rectangle, it follows that each of them is half of the area of the rectangle. For this to work, some triangles need to be cut into several pieces, as we will see. You will need to print the following (or make paper equivalents):

- **TriangleArea.scad** (TWO SETS needed)
  - This model consists of a one- or multi-part triangle. It has an option to print out one or two triangle sets. Set the variable **twosets** to TRUE to make two sets at once, or FALSE to make one (if, for example, you want to print them in different colors, as we have done here).
- **TriangleBox.scad**
  - This model creates a rectangular open box of the same base and height as the TriangleArea triangles.
- **TriangleAreaBox.scad**
  - This model creates a triangular open box that holds together just one TriangleArea triangle

These models all have some combination of the following variables:

- **base = 105**
  - the base of the triangle, in mm
- **height = 70**
  - the height of the triangle, in mm
- **thickness = 20**
  - the thickness of the extruded triangle models, in mm
- **top = base * 1.3**
  - the position of the highest point of the triangle above the base
- **linewidth = 0.1**
  - If you set **thickness** = 0, the program will create a file you can export as a .svg file to print on paper. This variable is the width of lines drawn, in mm.

If you have to change the sizes of these triangles, it is better to change these numbers in OpenSCAD rather than to scale the models in a slicer because tolerances for parts to fit together may not work right if you scale everything up or down uniformly. Note that these models are designed to all work together and in some cases be fitted together. If you change one, you need to change them all.

If you change the position of the highest point (the value of the parameter **top**) you may get more or fewer pieces per triangle. (See the section "Paper Version of Triangle Area Model" to see the parameter values that will give you a version you can print on paper.)

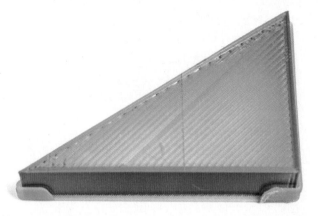

FIGURE 5-19: Right triangle in its box

We will walk through three examples of varying complexity to give you some ideas on how to use the model, from a very simple one through a bit of a puzzle.

RIGHT TRIANGLE MODEL

First, let's test whether two right triangles, similar to our folded square piece of paper, will indeed make a rectangle. First, set the parameter **top** in **TriangleArea.scad** and **TriangleAreaBox.scad** to:

```
top = base * 1.0;
```

This will create a right triangle, with its top vertex directly over the right-hand side of the model, as we see in Figure 5-19.

Then 3D print these models, leaving the other parameters at their defaults noted earlier:

- Two sets of the triangle from **TriangleArea.scad** (or you can do this by setting the parameter **twosets** to TRUE) to do both at once).
- One triangular box with **TriangleAreaBox.scad** (optional; we'll see why shortly)
- One of the rectangular boxes from **TriangleBox.scad**

This triangle prints in just one piece. In Figure 5-19 we see the triangle from **TriangleArea.scad** in the organizing box made by **TriangleAreaBox.scad.** The box isn't really useful here, but we show it here for consistency with the more complex cases we will do in the next section.

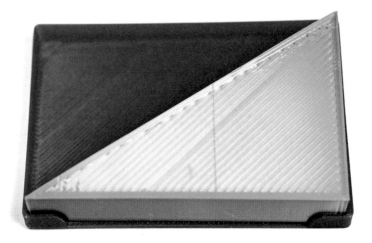

FIGURE 5-20: One triangle in a box with the same base and height

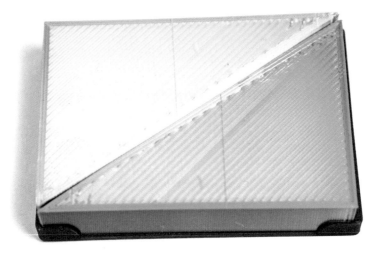

FIGURE 5-21: Two triangles fitting in the same box

Next, we created an open-top rectangular box with the same base and height as our triangle, using TriangleAreaBox.scad. In Figure 5-20 we see one green triangle model in the blue box. In Figure 5-21, we see that if we add a second, identical triangle, they exactly fill the box. (The boxes are slightly bigger than the triangle pieces so they are easy to take in and out.)

ACUTE TRIANGLE MODEL

Now, let's see if we can do the same thing with an acute triangle. Set the parameter **top** in **TriangleArea.scad** and **TriangleAreaBox.scad** to:

FIGURE 5-22: Triangular box, showing the overall dimensions of the triangle.

FIGURE 5-23: Triangle pieces in their box

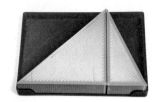

FIGURE 5-24: One triangle in the rectangular box

```
top = base * 0.7;
```

This will create a triangle with its top vertex over a point 70% of the way along its base.

Then 3D print these models, leaving the other parameters at their defaults noted earlier:

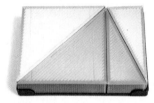

FIGURE 5-25: Two copies of the triangle exactly filling the rectangular box

- Two sets of the triangle from **TriangleArea.scad** (or you can do this by setting the parameter **twosets** to TRUE) to do both at once).
- One triangular box with **TriangleAreaBox.scad**

You can use the same rectangular box for this case as you did for the previous example if you keep the base and height parameters the same. **TriangleArea.scad** will print this triangle in two pieces. In Figure 5-22 we see the triangular box from **TriangleAreaBox.scad.** Figure 5-23 shows the two pieces of this triangle in that box. You can use the organizing box as a reminder of the shape of the original triangle.

Now try putting this triangle in the rectangular box. If you did not change the base and height from the last example, you can use the same box.

Now, if indeed the area of this triangle is also half its base times its height, we should be able to fit another one of these triangles in the rectangular box. However, it is obvious that the second set will need to be cut into pieces and arranged to fit. We can see how that worked in Figures 5-24 and 5-25. As we expected, two copies of this triangle fit exactly into a rectangle of the same base and height.

OBTUSE TRIANGLE MODEL

Now, let's try this with an obtuse triangle whose top vertex is shifted off to

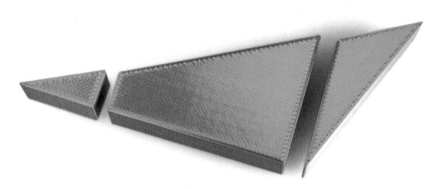

FIGURE 5-26: The obtuse triangle with the top vertex extending out over a point beyond the end of the base.

FIGURE 5-27: The obtuse triangle box that can hold the triangle in Figure 5-26

one side of the base. Set the **top** parameter back to its default value:

```
top = base * 1.3;
```

and make two copies of the **TriangleArea.scad** model and one of the **TriangleBox.scad** model. **TriangleArea.scad** breaks this triangle into three pieces (Figure 5-26). The triangular box created by **TriangleBox.scad** once again can remind us what the original triangle looked like as we try to fit it into our base * height rectangle (Figure 5-27). (You can use the same rectangular box as the previous two cases (Figure 5-28).

Now, take the three triangle pieces and arrange them in the triangle-shaped box so that you have a triangle that is easy to handle. There is one long side, one short side, and one side that is in-between. Arrange it so that the middle-length side is nearest you, and the shortest side is on the right (Figure 5-29).

FIGURE 5-28: Our rectangular box, the same for all three cases.

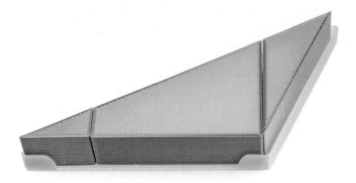

FIGURE 5-29: The obtuse triangle fitted into its box.

We call the side nearest you the *base* of the triangle, and the distance straight up to the highest point of the triangle (not the length of the side) is called the *height*. Note that we could turn the triangle around any way we wanted to, but to match the rest of the models for this discussion we need to orient it this way. If we had made the longest side the base, it would have fit into the rectangular box in only two pieces. It would have had to be a shorter, wider rectangle with the same area, rather than the same one we've been using.

Next, take the rectangular box. Line up its longest side with the base of the triangle. You'll see they are the same length (Figure 5-30). Then you can do

FIGURE 5-30: Showing the base of the rectangle and this triangle are the same.

FIGURE 5-31: Showing the height of the rectangle and this triangle are the same.

the same thing on the side to show that the other side of the box is the same as the height (Figure 5-31).

Line up the height (the vertical distance from the base of the triangle to the highest point) with the shorter side of the rectangular box. You can see that the side of the rectangular box is the same as that of the height of the triangle. The area of the box is the base times the height.

Now, this time around it takes a bit more fussing around with the parts of two copies of this triangle to find an arrangement that will fit into the rectangular box. Because the top point of this triangle extended beyond the base, it needs to be cut into three pieces to fit into the box. Figure 5-32 is one of the several solutions that work.

Now we have seen all three possible positions of the top vertex: directly over one of the other vertices, somewhere mid-base, and skewed over beyond the base. Now we know for all these cases:

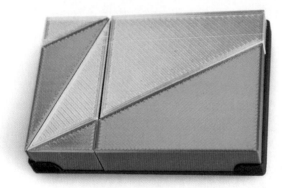

FIGURE 5-32: Two copies of the triangle fit into the rectangular box with the same base and height.

Area of a triangle = 1/2 * base * height.

If the **top** variable is greater than **base * 2**, the triangle will need to be cut into even more pieces, but it will still be possible to fit two copies into the rectangle. Create yourself a few more cases, either with 3D prints or paper, to convince yourself this works in general.

Now notice that in each of the examples, we broke the triangles apart so that at least one of the segments was based on a right angle. A right angle is necessary for the triangles to fit into a rectangular box. Now here, we're measuring the base * the height. What is that centered around? A right angle! So you can see that when we are creating these models, we are rearranging the parts of the triangle so that we find the triangle's height, which equals the side of our rectangle. We cut the base into pieces, but as you can see the base of the green triangle remains the same as the base of the rectangle. We rearrange the other pieces to fill the space above the triangle's base, up to its original height.

PAPER VERSION OF TRIANGLE AREA MODEL

If you are using paper instead of 3D printing, you can use **TriangleArea.scad** to make a file that you can print on paper. Open TriangleArea.scad in OpenSCAD. Set the variable **thickness** = 0. As described in Chapter 2, render the model. Use the **File > Export** option to export the file in .svg format.

If you want the lines to be a bit thicker, increase the value of the **linewidth** variable. The output will look like this for the first case (Figure 5-33, same

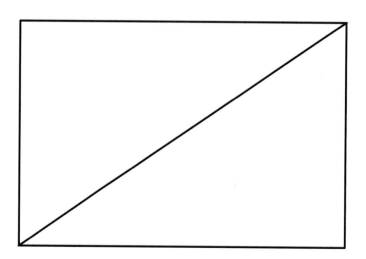

FIGURE 5-33: The equivalent of Figure 5-20 to be printed on paper

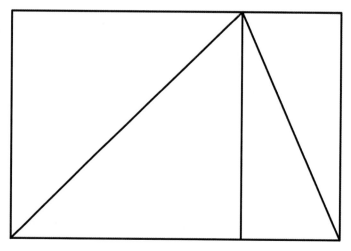

FIGURE 5-34: The equivalent of Figure 5-24 to be printed on paper

as the print in Figure 5-21), the second case (Figure 5-34, same as the print in Figure 5-25)) and the final case (Figure 5-35, same as the print in Figure 5-32).You can print out a few copies, and use one as a template for the finished model. This type of file can be dragged into a browser, Adobe Illustrator, and other programs and printed in hardcopy from there. It can also be used for laser cutting. You could try just cutting the triangles yourself until they fit, but the cuts need to be in certain places to fit with the smallest number of pieces, and the rules for determining where to cut are encoded in the OpenSCAD code.

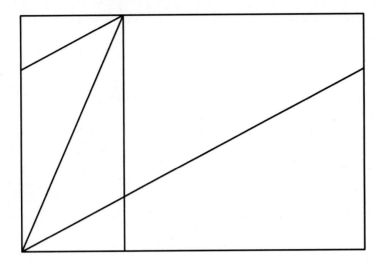

FIGURE 5-35: The equivalent of Figure 5-32 to be printed on paper

WHY TRIANGLES MAKE THINGS STRONG

Things made up of triangles can be very strong. You can, in fact, see them in the wild as parts of bridges, houses, and just about any mechanical structure you can imagine. A truss is a structure made up of triangles that appears in many constructions, which we will learn about in detail in Chapter 13. We can demonstrate the relative strength of triangles and squares with a few drinking straws and some elastic cord.

First, take two drinking straws. Plastic will work better, but sturdy paper should be okay as well. Cut four equal-length pieces of straw. If your straws are bendy straws, be sure not to use the bendy part. Run the elastic (0.2mm works well) through all four pieces, and tie off the cord (Figure 5-36). You can tuck the knot inside one of the straws to hide it. Now you'll have something that looks like this.

Notice that, since the corners are free to rotate, it will not really want to maintain a square shape at all or even stay flat, particularly if you squash it at all.

Now take another two straws, again cut off the bendy part, and use three of the parts to make a triangle with an elastic cord running through (Figure 5-37). Tie the cord off and tuck the knot into one of the straws so all the joints move freely. Now, try to squash it by pressing on one of the points while

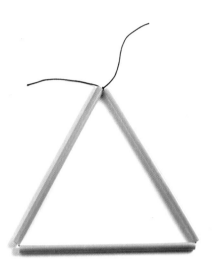

FIGURE 5-36: Straws and elastic can't maintain a rectangle.

FIGURE 5-37: A sturdy triangle, also made of straws and elastic

holding onto the opposite side. What happens?

Play with the square and the triangle. You'll see that the triangle shape can have its joints free to move, and yet be very stable. Think about why this is and test out your hypothesis. Head to Chapter 13's discussion of trusses to see if you were right!

You also can try adding a cross-brace to the square with another drinking straw and some more cord. Tie the cross-brace in place at the corners (Figure 5-38). Remember that this diagonal piece has to be longer than the sides, since it is the hypotenuse of what will be two right triangles. What do you expect to happen? Again, we will get into this more in Chapter 13.

3D PRINTING MESHES

Now that we've seen a lot of the inner workings of triangles, let's see if we can observe some of them in the wild. One of the reasons that triangles are interesting is that you can break any other shape with straight sides into triangles by strategically drawing some lines. In fact, if the triangles get small enough, you can even start to approximate curved surfaces by covering them with tinier and tinier triangles. Imagine that you had a large number of triangles ranging in size from a grain of rice to a few inches on a side, and that you wanted to cover a heart-shaped pillow. With a lot of fussing around, you could sew those triangles together to cover the pillow.

FIGURE 5-38: A rectangle with cross-bracing

That is, loosely speaking, how software like OpenSCAD makes an "STL" (standard tessellation language) file that models a complicated surface. "Tessellation" means to cover a surface with a geometrical shape (such as triangles). STL files consist of long lists of the three vertices of triangles, and how exactly that triangle is rotated in space. If you get more involved with 3D printing, you will hear people talking about these *meshes* of triangles, and worrying about various problems that can make them hard to print.

For example, if a mesh has some triangles missing (like a place in our hypothetical pillow, not covered by cloth, with stuffing coming out) that is said not to be *watertight* (since the hole would let out water). If the surface folds in on itself in a way that will not make sense in the physical world (two parts of the surface passing through each other, for instance) we say the mesh is not *manifold*.

3D printer software (like Ultimaker Cura or PrusaSlicer) takes these STL files and figures out what commands need to be sent to the printer to create the object layer by layer. These programs are now pretty good at figuring out the user's intent when a mesh isn't watertight and manifold, and you can get away without worrying about it for the most part. But, while we are talking

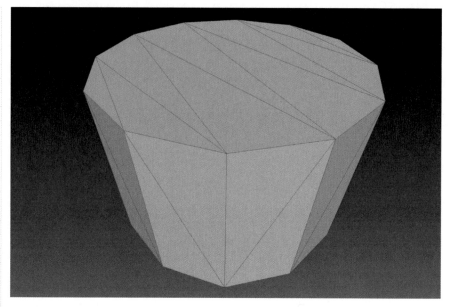

FIGURE 5-39: Screen capture from Meshlab, showing triangles

out these issues, you might want to explore the 3D printing mesh that makes up an STL file. You can do that with Meshlab, which is a bit challenging to use, but very powerful.

VIEWING A 3D PRINT'S MESH

The open-source program Meshlab (www.meshlab.net) allows you to view your meshes. For example, this is what the mesh looks like for a 12-sided prism model (Figure 5-39). You can see that the curved surface required a lot of little skinny triangles to approximate a curve. We will see more of this sort of approximation in Chapter 7.

Meshlab is pretty complicated, but often if you just open an STL file in Meshlab, it will detect problems and offer to fix them for you. Typically it does a good job, and you re-export your mesh, and then use it in 3D printing software as normal. You have to be a little careful with Meshlab, though, since it will overwrite the STL file you gave it with one it thinks is better.

You might try installing Meshlab and viewing one of the model meshes. All you need to do is to drag a .stl file into Meshlab. Then click the "Wireframe" icon (it looks like a little birdcage) to see the triangles(Figure 5-40).

FIGURE 5-40: Icon to click in Meshlab to see surface triangles in an STL file

SUMMARY AND LEARNING MORE

In this chapter, we learned about the many types of triangles, and how to recognize them in the wild by their angles and relationships among their sides. Wikipedia has a lot of good information: the entry *Triangles* is a good place to find lots of links to more information.

Later in this book, we have two chapters that have larger projects that tie together several ideas in the book. We'll give you more background in the intervening chapters, but you might want to peek ahead to see where we are going. In Chapter 7, we will have you use the ideas from this chapter, plus more about circles to find your latitude. Chapter 12 will build on Chapter 7 to show you how to know even more about finding out what time it is based on simple geometry!

If you look at bridges and buildings, you see triangles everywhere you turn. In Chapter 13, we will talk about structures made of triangles (and tetrahedrons) called trusses, and think about why they are so strong.

Next up we are going to learn about the Pythagorean Theorem, which shows us how the sides of a right triangle relate to each other. It turns out that this ancient and simple relationship underlies modern calculations in applications from navigation to architecture. Then, we will move into a quick introduction to the trigonometry we need to be able to understand some of the cooler geometry problems we explore in later chapters.

ANSWERS

Here are the answers for the activities in this chapter that we don't solve in the text of the section.

TRIANGLE-TYPE TABLE

	Acute	Obtuse	Right
Equilateral	YES	NO	NO
Isosceles	YES	YES	YES
Scalene	YES	YES	YES

Equilateral triangles can't be right or obtuse, because all of their angles are 60°, which makes them acute angles. Acute, obtuse, and right triangles can all be isosceles because regardless of how wide their largest angle is, the other two angles can still be equal to each other. Likewise, a scalene triangle's widest angle can be acute, right, or obtuse.

CHAPTER 6
PYTHAGORAS AND A LITTLE TRIGONOMETRY

You may be detecting that a lot of what we have talked about has its roots in Greece somewhat more than 2000 years ago. We'll keep that theme in this chapter, starting with a theorem by Pythagoras, a Greek philosopher who lived around 2500 years ago and is known for advocating common living and sharing of goods as well as for mathematical ideas. The record is a little murky that long ago, but many historians think he was the one to come up with a theorem that bears his name, about how the lengths of the sides of a right triangle relate to each other. It seems a little magic, or a coincidence, but we will prove it, and use it, in the early part of this chapter.

Then, armed with new ideas about right triangles, we will meet the Greek astronomer Hipparchus, who lived about 2100 years ago and who is regarded as the father of trigonometry. Even though people talk about the difference between a geometry class and a trigonometry class, in the end it is all math, providing you with tools that you can use to solve problems. In that spirit, this chapter gives you a bit of trigonometry (trig, to its friends) that we'll use in the chapters and applications that follow.

We hope you are also feeling how it must have been for these long-ago mathematicians when they realized they were understanding something for the first time ever. We love how it feels when another concept clicks together in our heads; mathematicians often use the word "elegant." To be a little more pragmatic, think of it as another rung to snap on a ladder you're building while you climb it - be sure each piece is firmly attached before you rush upwards!

3D Printable Models Used in this Chapter

See Chapter 2 for directions on where and how to download these models.

ExtrudedTriangle.scad
Just two lines long, this model draws a triangle from any three points and extrudes it to the desired thickness.

theodorus.scad
Creates a Spiral of Theodorus

hypotenuse.scad
Creates a "slide rule" triangle model for measuring sine and cosine.

Other supplies for this chapter
- 25 2x2 LEGO bricks
- A few pieces of construction or similar heavy paper
- Regular paper to sketch on
- A calculator that can find sine, cosine, and tangent
- Ruler

PYTHAGOREAN THEOREM

About 2500 years ago, Pythagoras figured out that if you have a triangle that has one right angle, there is a relationship between the longest side and the two shorter ones. Specifically, if we add the squares of the two shorter sides, we get the square of the longest side (the hypotenuse). As an equation, if the two shorter sides are of length a and b and the longest is of length c, the Pythagorean Theorem says:

$$a^2 + b^2 = c^2$$

For example, if the hypotenuse of a triangle is 30mm long, then 30mm squared, or 900 sq mm, is the sum of the squares of the sides. Let's say that we want the other two sides to be the same length as each other(a = b). We know that

$a^2 + b^2 = c^2$
Since a = b, then $(2 * a^2) = c^2$
$2 * a^2 = 900$
$a^2 = 900/2$
now take the square root of a^2
a = b = 21.2mm

Squares and Square Roots

If symbols like a^2 are new to you, they may look scary but are pretty easy to understand. What a^2 means is "take the number a and multiply it by itself." Or, in other words, a^2 is a * a. Then if we want to multiply by a one more time, we write $a^3 =$ a * a * a, and so on.

We can go the other direction, too. Suppose we know that $a^2 = 2$. That means that something times itself equals 2. Figuring this out is called taking a square root, and its symbol looks like this: $\sqrt{2}$
You probably have a button on your calculator that says "$\sqrt{}$", and if you want to use Google to get the square root of 2, you would type sqrt(2).

PYTHAGORAS' LEGO

We can demonstrate this with LEGO bricks and a 3D-printed triangle. A 2 * 2 LEGO brick is 16mm * 16mm by just shy of 10mm deep. If we have four of them on one side (64mm), and three on the other (48mm), that means that the square of the longest side should be 3 squared plus 4 squared, or 9 + 16 = 25 (Figure 6-1). The hypotenuse of the triangle should fit 5 bricks exactly (Figure 6-2).

You can use the model **ExtrudedTriangle.SCAD** to generate a triangle with one side 64mm long and the other 48mm long, at right angles to each other. We will want it to be 10mm thick. To do that, update ExtrudedTriangle.scad as follows. Or just use these two lines as your model; see Chapter 2 to learn how to make a new OpenSCAD model.

```
thickness = 10;
```

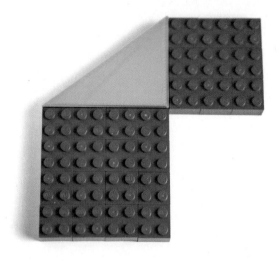

FIGURE 6-1: The sides of the triangle line up with 3 and 4 bricks each.

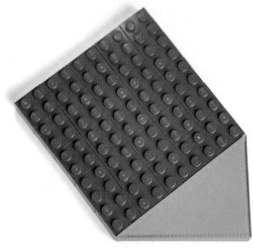

FIGURE 6-2: The hypotenuse of the triangle lines up with 5 bricks.

```
linear_extrude(thickness) polygon([[0, 0], [64, 0], [64, 48]]);
```

Take that triangle and 25 2 * 2 LEGO bricks. Make a square of 9 of the bricks and a square of 16 of them. These will align with the two short sides. Now rearrange these bricks into a 5 by 5 square, which will line up with the hypotenuse.

If you use a different-size counting piece, you will need to adjust the triangle you create. There are a few "Pythagorean Triples" that come out to nice round numbers like this (the next one is 5, 12, 13), and there are long lists available if you search on that phrase.

You can use the Pythagorean theorem any time it is easy to measure two distances or lengths in directions perpendicular to each other, and you want to know the shortest distance between the farthest-apart points of the triangle. If you started at the left end of the hypotenuse and traveled right and then up to get to the right end, your path would be longer than if you just traveled along the hypotenuse.

MAKING A SPIRAL OF THEODORUS

What happens if we construct a series of right triangles in succession? About 2500 years ago, Theodorus of Cyrene came up with a way to make a physical calculator, of sorts, for square roots based on the Pythagorean theorem. Theodorus probably lived a bit later than Pythagoras, but his life is less well-documented. His construct is called the Spiral of Theodorus (or, sometimes,

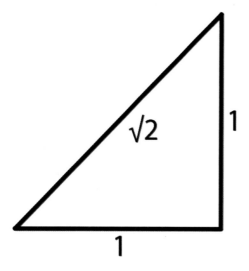

FIGURE 6-3: First triangle for Spiral of Theodorus.

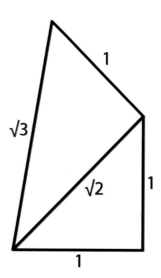

FIGURE 6-4: First two triangles for Spiral of Theodorus.

FIGURE 6-5: 3D print of first three triangles for Spiral of Theodorus.

FIGURE 6-6: 3D print of first 17 triangles for Spiral of Theodorus.

the Pythagorean Spiral, but we'll give the lesser-known mathematician his due here).

First, let's make a Spiral of Theodorus with just a ruler and paper, so we can see where things came from. First, draw a right triangle with the two sides equal to 1cm in length. (Units don't matter, since everything is a ratio later on, but it might help to try this exercise on graph paper.) When you draw in its hypotenuse, the Pythagorean Theorem says its length will be the square root of $1^2 + 1^2 = 1 + 1$, or $\sqrt{2}$ (Figure 6-3).

If we wanted to get $\sqrt{2}$ without a calculator, we could just measure that side in centimeters (if our other sides were both 1cm). But Theodotus had a bigger idea. What if you kept going, and added another triangle with its sides equal to 1 and $\sqrt{2}$? Then the hypotenuse of THAT triangle would be $\sqrt{3}$ (Figure 6-4).

FIGURE 6-7: 3D print of first 53 triangles for Spiral of Theodorus.

FIGURE 6-8: 3D print of first 150 triangles for Spiral of Theodorus.

You can keep going as long as you want. The figure comes back around to the original point at about $\sqrt{17}$. We created an OpenSCAD program called **theodorus.scad** to let you create as big a spiral as you want (and can fit on your 3D printer). Here is a spiral with 3 triangles (Figure 6-5), which ends at a triangle with a hypotenuse of $\sqrt{4} = 2$.

Here is a spiral with 16 triangles (Figure 6-6, longest hypotenuse $\sqrt{17}$). The model creates a solid base for the figure for easier printing.

In principle, you can measure off the square roots of numbers from 1 to 17 from the flat version, which might have been handy in the era before calculators. The OpenSCAD model allows you to go beyond 16 triangles by wrapping the bigger triangles around underneath the first 16. Here are the models for 53 (Figure 6-7) and 150 (Figure 6-8) triangles.

You can do this with numbers that are not integers, but it is a bit messier. If we started with a triangle that has sides equal to 0.5 and 1 instead of 1 and 1, the hypotenuse would be $\sqrt{1.25}$. The next triangle would be $\sqrt{1.25}$ - 1 - $\sqrt{2.25}$, and so on.

The model theodorus.scad has a few variables you can work with:
- **max** = number of triangles to be made (note that the longest hypotenuse will be $\sqrt{max + 1}$
- **xyscale** = length of the "equal to 1" sides of the triangle, in mm
- **zstep** = the vertical step for each triangle (in mm, but needs to be a multiple of layer height)

If you want to read square roots off the model, you'll need to measure the length of the lines from the center carefully using a ruler or, preferably, calipers. You also will need to allow for line width (defaulted to 1 mm). In essence, we are drawing the triangles with a pencil 1 mm thick. The lines are centered on the desired value, so measuring the outside of each triangle will give you a value that is too large by one wall width. You should correct for that by subtracting the wall variable from your measurements in mm.

If you want to draw a set with a pencil instead, and measure off the hypotenuses, be sure to measure the first (and subsequent) 1 cm sides carefully and be sure that the right angles are really square. If you draw the triangles larger, and using a finer pencil line, you'll be able to read the lengths more precisely. You can draw the triangles as large as you want. Just remember that the hypotenuse will be scaled up by the length of a side of the first triangle. For example, if that is 3 cm, you would have to divide the length of each hypotenuse by 3 cm to get the square root.

A LITTLE TRIGONOMETRY

No parts of mathematics are really completely separate from the other. In particular, geometry and trigonometry have some common roots that we need to comprehend if we're going to understand both subjects. In this section, we will introduce a bit of trigonometry that describes useful relationships between the sides and angles of right triangles. First, we'll define these relationships, and then we'll put them through their paces and see how they can help us out.

SINE, COSINE, TANGENT

It turns out that there are ratios among the sides of right triangles that can be used to calculate angles, or other sides of the triangles. Three ratios are used most commonly, called *sine*, *cosine*, and *tangent*. These deal with the ratios of the triangle's three sides. The hypotenuse is one of these sides, and the other two are referred to as the *adjacent* side (the side sharing a particular angle with the hypotenuse) and the *opposite* side (the third side of the triangle, which does not share that angle).

The *sine* of an angle *a* (usually written sin(a)) in a right triangle is

sin(a) = opposite divided by the hypotenuse

The *cosine* of an angle *a* (usually written cos(a)) is

cos(a) = adjacent divided by the hypotenuse

FIGURE 6-9: The hypotenuse model assembled

FIGURE 6-10: Labeling on the bottom and side of the model

The tangent of an angle a (usually written tan(*a*)) is

tan(a) = opposite divided by adjacent, or sin(a)/cos(a)

HYPOTENUSE MODEL

We will use the model **hypotenuse.scad** to get some intuition about these ratios. First, assemble the model. It has two legs; one has a slot cut through it, and one has a slot cut part-way down and is solid at the base. Place the model so that the leg that is solid at the base is on the right side, and the one with the slot all the way through is on the bottom. It will look like a backward letter L. Next, take the slider and put one of its tabs into each slot. The slider plus the L-shaped part makes a right triangle (Figure 6-9). You now have an analog calculator of sorts for sine and cosine.

We will talk about the angle on the bottom left of the triangle first. Let's say that the hypotenuse of the triangle made by the slider is one unit long. The bottom and side of the L shape are marked off in tenths of the hypotenuse (Figure 6-10).

Therefore, since sine is equal to opposite/hypotenuse, we can read off the sine of this angle on the right side, and cosine off the bottom (adjacent) side. The zero point is at the lower right (where the two legs join) and the tenths are marked with long lines with nine lines in-between. Thus you should be able to read the value of the sine or cosine to two digits. The lines are indented for better printing and longer model life.

FIGURE 6-11: 0° shown on the model

FIGURE 6-12: 90° shown on the model.

Similarly, you can read off the cosine of the angle from the scale along the bottom of the model (the side cut all the way through). You will read it *right to left* — that is, zero is at the corner for *both* scales. Figure 6-10 shows a cosine reading of about 0.62.

Since the hypotenuse is the longest side of the triangle, the value of sine goes from 0 to 1. You can see this by sliding the slider all the way over one way till it is nearly flat (near 0°, as shown in Figure 6-11) to the other end where it is nearly vertical (90°, as shown in Figure 6-12).

Finally, what about the tangent of this angle? Tangent is the opposite side over the adjacent, or sine divided by cosine. This means it is how much the slider rises from left to right versus how far it is from the corner. When we learned about lines in coordinate systems in Chapter 2, we saw that this is called the *slope* of the line made by this angle.

Unlike sine and cosine, there is no limit to the tangent of an angle. As the angle approaches 90°, the tangent approaches infinity, because the bottom of the fraction opposite/adjacent is approaching zero. Two angles of a triangle can't actually both be 90°, though, since the angles need to add up to 180°. That is consistent with the fact that dividing by zero is also undefined.

Test this out. Use the model to find what angle will give you a sine of about 0.71. What is the cosine of this angle? Describe what happens to the sine, cosine, and tangent of an angle when the angle gets close to zero. What about as it gets close to 90°? Use the model to help demonstrate and think about this. (See the answers at the end of the chapter for the numbers.)

USING A TRIANGLE AS AN ANGLE TEMPLATE

If you want to set the angle more precisely, you can create a triangle with **Extruded_Triangle. scad** or set from **TriangleSolver.scad** as an angle template. Your triangle template does not have to be a right triangle, since you only care about the one angle you are measuring.

COMPLEMENTARY ANGLES

We know that the other two angles in a right triangle add up to 90°. If we use the model and think about the two angles, the sine of one angle is the cosine of the other. The "co" in cosine stands for "complementary." The sine of an angle is the same as the cosine of its complementary angle. Or to put it another way, $\sin(x) = \cos(90° - x)$ and $\cos(x) = \sin(90° - x)$.

Along with complementary angles, you may hear the term *supplementary* angles. This refers to two angles that add up to 180°. A pair of supplementary angles will have the same sine, and their cosines will add up to zero, as we'll see when we talk about angles between 90 and 180.

OTHER RATIOS

Sometimes it is more convenient to use other trigonometric ratios, which we will define for you here.

secant (sec) = 1/cosine
cosecant (csc) = 1/sine
cotangent (cot) = 1/tangent

Like sine and cosine, the other complementary (co-) ratios have this relationship with the complementary angles:

FIGURE 6-13: Equilateral triangle as a template

Degrees, Radians, and Pi

We have been talking about angles in degrees to this point in this book. As we explored in Chapter 5, a degree is a measurement of how big an angle is. 360° makes a whole circle; a rotation of 180° points us in the opposite direction (half a circle). 360° was chosen (most likely by the ancient Babylonians) because it is easy to divide into halves, thirds, fifths, etc.

Sometimes it is more useful to be able to think about fractions of a circle in terms of the fraction along the circle's perimeter (or circumference) you have traversed. The unit of that measurement is the radian, which is based solely on values in nature, like pi. 2π radians take us around a whole circle. Imagine we have a circle with a radius = 1 in some units. The circumference of that circle would be $2\pi r$, where r is the radius.

Since the radius is 1, the circumference is just 2π. If we had a wedge of the circle with a 45° angle at the center of the circle, it will have gone through 45/360ths of a circle, which translates to 1/8th of a circle. If the circumference is 2π, then $2\pi/8 = \pi/4$. An angle of 45° is thus the same as an angle of $\pi/4$ radians. 90° is $\pi/2$ radians, and 180° is π. And of course, 360° is 2π radians.

In general, the formula to convert is:

radians = degrees $*2\pi/360°$, or, more simply, radians = degrees $*\pi/180°$.

Or degrees = radians $* 180°/\pi$.

If you use a calculator or computer program to compute the ratios in this chapter and beyond, be sure you know whether it is using degrees or radians. OpenSCAD uses degrees, but most other computer languages use radians. Calculators that include trigonometric functions usually let you switch between degree and radian modes. Google's calculator uses radians.

Just to make things more interesting, sometimes degrees are shown as decimal degrees, like 34.1028°. Other times, they are shown in degrees, minutes, and seconds (written like this: 34° 6' 10", which I would read 34 degrees, 6 minutes, 10 seconds).

To go from degrees-minutes-seconds to fractional degrees, divide minutes by 60 and seconds by 3600, and add the result to the number of degrees. Example: 34° 6' 10" is 34 + 6/60 + 10/3600 = 34.1028°.

To go the other way (from decimal degrees to minutes and seconds) multiply the decimal part by 60 to get minutes. For example, if we have 34.1028° and multiply the 0.1 by 60, that's 6.1667 minutes. If there was a fractional minute, we would multiply that fraction by 60 again, in this case, 0.1667 times 60, or 10 seconds.

$$\tan(x) = \cot(90° - x)$$
$$\sec(x) = \csc(90° - x)$$

One tricky thing though is that because cosine, sine, and tangent can all go to zero, secant, cosecant, and cotangent will approach infinity when those numbers in the denominator go to zero. (A mathematician would say that dividing by zero is *undefined*, and these ratios are said to be *indeterminate* as they approach infinity.)

FINDING THE LENGTH OF A SIDE

Let's suppose that we know that an angle of a right triangle is 30°, and the hypotenuse is 5cm long. What is the length of the opposite side? We can use a calculator (or estimate with our model) to find out that the sine of 30° is 0.5. This means that the opposite side is 0.5 times as long as the hypotenuse, or 2.5cm.

$$\sin(30°) = 0.5 = \text{opposite/hypotenuse} = \text{opposite/5cm}$$

Do the same thing with cosine to find the adjacent side.
$$\cos(30°) = 0.866 = \text{opposite/hypotenuse} = \text{opposite/5cm}$$
$$\text{Opposite side} = 0.866 \text{ of the hypotenuse} = 0.866 * 5 = 4.33$$

Note that you can check that you are right because the Pythagorean Theorem says that the squares of the two sides should equal the square of the hypotenuse. In our case here, is it true that:

$$2.5^2 + 4.33^2 = 5^2$$
$$6.25 + 18.75 = 25$$
$$25 = 25, \text{ so our answer is correct.}$$

CALCULATING WITH SINE AND COSINE

You can test out how well you understood the discussion above by trying to do some calculations yourself. The answers are at the end of the chapter.

- I have an angle of 45° in a triangle with a hypotenuse 5cm long. What is the length of the opposite side?
- What is the length of the adjacent side? Why? (Hint: if one angle is 45°, what is the other angle in a right triangle?)
- Check to see that you are right by using the Pythagorean Theorem.

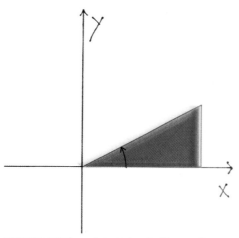

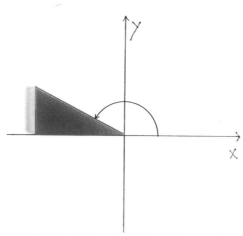

FIGURE 6-14: Our original triangle. The small arrow marks the angle, A.

FIGURE 6-15: Triangle from 6-14 reflected across the y-axis

ANGLES NOT BETWEEN 0° AND 90°

What happens if you are trying to take the sine or cosine of an angle larger than 90° (or negative angles)? The answer is that the definition gets a little more complicated, but not by a lot. First, let's remember that in Chapter 2 we learned about Cartesian coordinate systems, and in Chapter 3, about rotation and reflection (mirroring, in OpenSCAD).

We've printed out a plastic triangle that we can rotate and flip over. Let's put *x-y* coordinate axes behind our triangle. We are interested in the angle at the vertex of the triangle that is at the center of the coordinate system, which we will call "A". We've marked it with a small black arrow (Figure 6-14).

To find the sine of that angle, we need the height of the triangle or the side parallel to the *y* axis. We'll call that length *y*. To get the sine, we would divide *y* by the hypotenuse. Similarly, the cosine would be the length in the *x* direction, *x*, divided by the hypotenuse.

Pythagoras would tell us that the hypotenuse would be $\sqrt{x^2+y^2}$, so our formulas for sine, cosine, and tangent become:

$$\sin(A) = y / \sqrt{x^2+y^2}$$
$$\cos(A) = x / \sqrt{x^2+y^2}$$
$$\tan(A) = y / x$$

So far, except for showing Figure 6-14's plastic triangle on a coordinate

---

plane rather than as a triangle with its sides and angles labeled in isolation, this is the same situation for finding sine, cosine, and tangent as we've discussed earlier in this chapter.

ANGLES BETWEEN 90°AND 180°

Now, what would happen if we flipped our triangle over (which a mathematician would call "reflecting" it across the *y* axis). Let's suppose that now we wanted to know the sine, cosine, and tangent of the large angle marked by the arrow in Figure 6-15. This angle is 180°- A, since together they make a straight line. If we can figure out what the trigonometric ratios are for our flipped-over triangle, then maybe we can deduce what they are for the bigger angle.

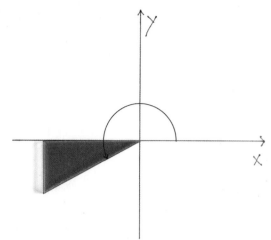

FIGURE 6-16: Flipping the original triangle in Figure 6-14 over about the y and then the x axes.

Flipping our original triangle over can be thought of as subtracting our angle A from 180°, because we have to back up angle A degrees from the straight line, which is 180° from our original starting point for measuring A.

The plastic triangle didn't change when we mirrored it, did it? So we know its angle A must be the same (in Figures 6-14 through 6-18 the angle A will stay the same since it is the same blue plastic triangle). However, the x direction now points in the other direction, so what was *x* now becomes *-x*. The y value, however, stays the same. The length of the hypotenuse stays the same. Also, x^2 and $(-x)^2$ are the same as each other, since a negative multiplied by a negative is a positive. Given all that, we get:

$$\sin(180° - A) = \sin(A) = y / \sqrt{x^2+y^2}$$
$$\cos(180° - A) = -\cos(A) = -x / \sqrt{x^2+y^2}$$
$$\tan(180° - A) = -\tan(A) = y / (-x)$$

Therefore, to figure out the sine, cosine, and tangent of angles between 90° and 180°, we subtract the angle from 180° and adjust the signs accordingly. For example:

$$\sin(120°) = \sin(180° - 120°) = \sin(60°) = 0.866$$
$$\cos(150°) = -\cos(180° - 150°) = -\cos(30°) = -0.866$$
$$\tan(180°) = -\tan(180° - 180°) = \tan(0°) = 0$$

ANGLES BETWEEN 180 °AND 270°

Suppose now we flipped the triangle over again. It's still the same triangle, but now both the x and y sides are pointing in the negative direction (Figure 6-16). Now we are going to see if we can find a different relationship between our original angle and the bigger angle shown in Figure 6-16.

In this case, both the *x* and *y* values are negative. We can think of this as rotating the original triangle to 180° plus A.

$$\sin(180° + A) = -\sin(A) = -y / \sqrt{x^2+y^2}$$
$$\cos(180° + A) = -\cos(A) = -x / \sqrt{x^2+y^2}$$
$$\tan(180° + A) = \tan(A) = (-y) / (-x) = y / x$$

To calculate the sine, cosine, and tangent of the large angle shown in in Figure 6-16 (an angle between 180° and 270°), we subtract 180° from the angle, find the sine, cosine, or tangent, and again adjust signs accordingly. Some examples:

$$\sin(240°) = -\sin(240° - 180°) = -\sin(60°) = -0.866$$
$$\cos(180°) = -\cos(180° - 180°) = -\cos(0°) = -1.000$$
$$\tan(250°) = \tan(250° - 180°) = \tan(70°) = 2.748$$

ANGLES BETWEEN 270° AND 360°(OR -90° AND 0°)

Finally, suppose we went back to the original position of the triangle (where both x and y were positive) and flipped it over (reflected it) as we can see in Figure 6-17.

Now x is positive, and y is negative.

$$\sin(360° - A) = -y / \sqrt{x^2+y^2}$$
$$\cos(360° - A) = x / \sqrt{x^2+y^2}$$
$$\tan(360° - A) = -y / x$$

To calculate the sine, cosine, and tangent of the large angle shown in in Figure 6-17 (an angle between 270° and 360°), we subtract the angle from 360°, find the sine, cosine, or tangent, and again adjust signs accordingly. Some examples:

$$\sin(300°) = -\sin(360° - 300°) = -\sin(60°) = -0.866$$
$$\cos(330°) = \cos(360° - 330°) = \cos(30°) = 0.866$$
$$\tan(290°) = \tan(360° - 290°) = -\tan(70°) = -2.748$$

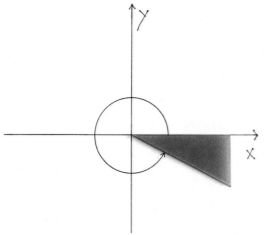

FIGURE 6-17: Our original triangle reflected across the x-axis

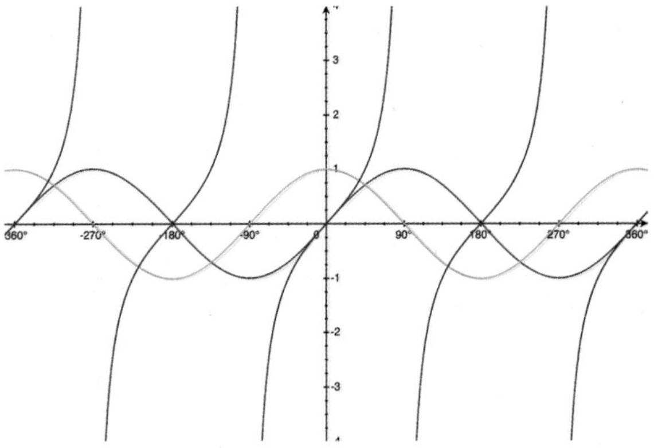

FIGURE 6-18: Sine(red), cosine(green) and tangent(blue) for angles from -360° to 360°

We can also equivalently think of this as going around the axis -A°. Negative angles move things around clockwise.

$$\sin(0° - A) = \sin(A)$$
$$\cos(0° - A) = -\cos(A)$$
$$\tan(0° - A) = -\tan(A)$$

Finally, if we kept going, and went around 360°, we would come right back to where we started. Adding 360° to any angle gives you all the same values for sine, cosine, tangent, and all the other ratios. This means that it is very handy to use these functions for things that are going around in circles, or that repeat somehow.

Now that we know how sine, cosine, and tangent work for angles more than 90 degrees, we can graph them for any angle. You can get values for these ratios either with a button on your calculator or by pasting something into Google for a calculation. However, Google expects angles in radians, not degrees, so you need to convert first, as we described in the sidebar "Degrees, Radians, and Pi" earlier in this chapter.

In Figure 6-18, we see the sine (red curve) and cosine(green) are gentle curves as the angle A increases (horizontal axis), with the signs of each of the ratios varying as shown, repeating the pattern each 360°. The tangent is shown in blue. Note that since tangent involves dividing by cosine, it approaches infinity when cosine approaches zero, at angles approaching 90° and 180°.

However, since the tangent is calculated as the opposite side divided by the adjacent side, as the angle goes to 90° the opposite side gets very big, and the adjacent side gets very small. As the adjacent side goes to zero, the tangent gets infinitely big (and ultimately is indeterminate, as we mentioned earlier in the chapter). The same thing happens, in the negative direction, at 270°. You can play with the hypotenuse model we showed you earlier in this section to prove this to yourself as the angle goes to 90°.

Try filling out the rest of this table with what the sign of sine, cosine, and tangent would be in each range of angles by looking back over this section. (Answer at the end of the chapter.)

	0° to 90°	90° to 180°	180° to 270°	270° to 0° (or -90° to 0°)
Sine	+			
Cosine	+			
Tangent	+			

ARCSIN, ARCCOS, ARCTAN

Suppose we wanted to "go backwards" and, given a sine, cosine, tangent or other ratio, find out what angle corresponds to it. In the case of sine of an angle A, this is called the arcsine (written asin( ), or sometimes arcsin( ) or inverse sine (written $\sin^{-1}$( )). The other ratios have inverses named similarly. For example $\sin(30°) = 0.5$, so $\mathrm{asin}(0.5) = 30°$; $\cos(60°) = 0.5$, so $\mathrm{acos}(0.5) = 60°$ and $\tan(26.6°) = 0.5$, so $\mathrm{atan}(0.5) = 26.6°$.

You might realize from the previous section that there are infinitely many angles that have the same values of sine, cosine, etc. since these functions repeat themselves (the official term is that they are *periodic*). Calculators will return the *principal value* of angles. For arcsine and arctangent, this ranges from -90° to 90°. For arccosine, the range is 0° to 180°. That way, there is no ambiguity about what the calculator is returning. Check the physical situation or any drawing you have to see if this is right. Also, be sure you know whether you are working in degrees or radians.

A good way to be sure you're doing something right is to take the sine of an angle, get the answer, and use a calculator to take the asin() of that answer. We did that a couple of paragraphs back when we showed how to go back and forth between sin(30°) and asin(0.5). Note that on some calculators, you need to hit another button (sort of like a shift key) to get to these functions.

SUMMARY AND LEARNING MORE

The Pythagorean Theorem is one of the best-known pieces of mathematics, and a little searching will give you more proofs if you would like to see it approached differently. To follow up on the trigonometry introduction, check out the *Trigonometric Functions* entry in Wikipedia. For more on sine (note that cosine and tangent only have links that redirect to the *Trigonometric Functions* Wikipedia entry) see the *Sine* entry. The Khan Academy has many videos in this space as well.

In the next chapter, you also are going to learn about circles. You will do that in part by trying out *constructions*, creating geometrical relationships with just a compass and straightedge. You'll also have enough pieces after that to use the ideas in this book so far to figure out your latitude just by observing shadows around lunchtime.

ANSWERS

Here are the answers for the activities in this chapter that we don't solve in the text of the section.

HYPOTENUSE MODEL

- Use the model to find what angle will give you a sine of about 0.71.
    - Answer: 45°
- What is the cosine of this angle?
    - Answer: also 45°- if one angle of a right triangle is 45° the other one has to be, too.
- Describe what happens to the sine, cosine, and tangent of an angle when the angle gets close to zero.
    - Answer: sine approaches zero, too (as the opposite side shrinks). Cosine approaches 1 as the adjacent side becomes about the same as the hypotenuse. Tangent approaches zero, since it is sine divided by cosine.
- What about as it gets close to 90°?
    - Answer: sine approaches 1 (as the opposite side and hypotenuse approach being equal lengths to each other.). Cosine approaches 0 as the adjacent side vanishes. Tangent approaches infinity, since it is sine divided by cosine.

CALCULATING WITH SINE AND COSINE

- I have an angle of 45° in a triangle with a hypotenuse 5cm long. Length of the opposite side = sin(45°) * 5cm = 0.7071 * 5cm = 3.54cm
- What is the length of the adjacent side? Also 3.54cm.
- Check to see that you are right by using the Pythagorean Theorem.
    - $3.54^2 + 3.54^2 = 25 = 5^2$ and so it works.

ANGLES GREATER THAN 90°

	0° to 90°	90° to 180°	180° to 270°	270° to 0° (or -90° to 0°)
Sine	+	+	-	-
Cosine	+	-	-	+
Tangent	+	-	+	-

CHAPTER 7
CIRCLES

Wheels, gears, soup cans, coins, and even the moon are some of the many things that are circles (or nearly so) in one of their dimensions. But in human history, we've also created circles that live only in our imaginations, like the circles that chop up the earth into handy measurements of east, west, north, and south.

Circles are handy when you are making something because you can draw them pretty easily with some string, or rope, or a gadget called a drawing compass (which we learned how to use in Chapter 4). In this chapter, we will tie together some of the material about triangles and trigonometry and use it to see what it was like to navigate with just the shadow of a stick and patience - and a lot of imagination.

AREA OF A CIRCLE

A *circle* is the set of all points that are a constant distance, the *radius*, from its center. You've probably heard that the area of a circle is pi times the radius squared. As we first saw in Chapter 3, pi is written as the Greek letter π, and is equal to about 3.14159. It is the ratio between the circumference, the distance one would travel to go all the way around a circle, and the *diameter*, which is twice the radius, or the longest distance across a circle.

We are going to try to figure out the area of a circle by breaking the area inside it into a lot of little triangles. In Chapter 5, we saw how to get the area of any triangle, and found out that it is just half the base times the height. In Chapter 3 we learned about regular polygons and saw that as a polygon had more and more sides it looked more and more like a circle.

Now we will tie some of that together in a very old proof that generates a regular polygon just

3D Printable Models Used in this Chapter

See Chapter 2 for directions on where and how to download these models.

inscribed.scad
Prints out a circle and its inscribed specified polygon

circumscribed.scad
Prints out a circle and its circumscribed scribed specified polygon

areaWedges.scad
Prints out wedges making up a circle, and a rectangular enclosure

gnomon.scad
Prints out a gnomon (which is the part of a sundial that casts a shadow) for measuring sun angle from the horizon

Other supplies for this chapter
- One toothpick
- Enough modeling clay or PlayDoh to make a ball about 2cm in diameter
- A ruler
- A lamp that can illuminate in all directions around it
- Optionally, an earth globe
- Cardboard (for alternatives to 3D printing)
- A calculator that can find sine, cosine, and tangent

FIGURE 7-1: Inscribed triangle (top left), square (top right), hexagon (bottom left), and dodecahedron (bottom right).

a bit bigger than a circle, then one just a bit smaller, to see if we can bracket the area of the circle that lies between that. (Remember that a regular polygon has all its sides the same length, and all the angles are the same.) Along the way, we can show where the formula for the area of a circle comes from, and why it works.

INSCRIBED AND CIRCUMSCRIBED POLYGONS

We are going to draw the biggest regular polygon we can inside a circle so that all the vertices of the polygon lie on the circle, as we can see in Figure 7-1 for a square, triangle, hexagon, and decagon. These are called *inscribed* polygons. We first met these in Chapter 3.

This also means that the circle is the smallest object that will fit around the polygon, so they are *circumscribed* circles. Regular polygons have a radius, the distance from their center to each vertex, which is the same as the radius of the circumscribed circle. (As we will see, we can do this the other way around, with a circle fitting inside a polygon. Then the circle will be inscribed, and the polygon, circumscribed.)

If we draw lines from the center of a regular polygon to each of the vertices, we break the polygon into a number of identical triangles, one for each side of the object. Here we see the triangles and angles associated with one side of each polygon. In the 3D printed model shown here, the angles are recessed so that you can feel the whole polygon shape as well.
These are *isosceles* triangles (two sides are the same), which we can split into a pair of identical *right* triangles, which will make it a little easier to find the height.

We know that the whole polygon is made up of a number of these right triangles equal to twice the number of sides. Way back in Chapter 3, we first saw that there are 360° in a circle. Taking those two facts together, the angle in the center for each of these triangles is 360°/(2 * number of sides). The base of each of these right triangles is half the length of one side. The triangles' height is the distance from the center of the side to the center of the polygon. This distance is called the polygon's *apothem*.

When the number of sides in a regular polygon gets really high, the sides get really short, and the polygon starts to look a lot like a circle.

By using larger and larger numbers of sides, we can see that the ratio of the polygon's area to the square of its radius gets closer and closer to pi as the shape gets closer and closer to a circle. (See the section "Bracketing Pi" later in this chapter to see the details.) As you can see in the model in Figure 7-1, as the polygons get more sides, the difference between the circle and its inscribed and circumscribed polygons has less and less area.

MODIFYING THE MODELS

The inscribed.scad and circumscribed.scad files have these variables that you can play with (defaults as noted).

- **wall_thickness** = 1.2
    - How thick to make walls that otherwise go to zero thickness, in mm. Note that ideally this variable should be set to twice your extrusion width. Defaults to 1.2.
- **radius** = 30;
    - circle (and polygon) radius, in mm
- **n** = 12;
    - polygon number of sides
- **height** = 10;
    - The thickness of these models, in mm

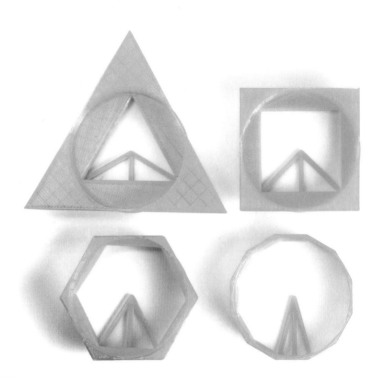

FIGURE 7-2: Circumscribed (blue) and inscribed (green) triangle (top left), square (top right), hexagon (bottom left), and dodecahedron (bottom right).

It is better to change these numbers in OpenSCAD rather than to scale the models in a slicer because tolerances for parts to fit together may not work right if you scale everything up or down uniformly. The inscribed and circumscribed models are designed to all work together. If you change these parameters in one model, you need to change them in both.

Try this experiment: print out the inscribed and circumscribed polygons for **n** = 4 (an inscribed and circumscribed square) and **n** = 12 (a dodecahedron). How many sides would it take for the polygon's area to be pretty much indistinguishable from a circle? In the sidebar "Bracketing Pi" we will calculate this example using sines and cosines, but see if you can estimate how far off each might be from looking at the models.

We can see that the printed parts of the models, which are the difference between the circle and inscribed or circumscribed polygons, get thinner as there are more sides. Laying the inscribed on top of the circumscribed polygons can give you this intuition (Figure 7-2).

Incidentally, you may notice that the models get very thin at the vertices of the inscribed and center of the polygon sides for the circumscribed models. In each case, we had to make the outer shape just a bit larger to maintain a minimum wall thickness. Otherwise, the models would get too thin to print in those areas, and the printer would create each model as several separate pieces.

BRACKETING PI

If you'd like to actually calculate how close you can get to the value of pi by looking at inscribed and circumscribed polygons, we'll work it out here. Let's assume we are going to approximate the area of a circle whose radius = 1 with an inscribed or circumscribed polygon with n sides. (If n = 3, for example, the polygon would be a triangle.)

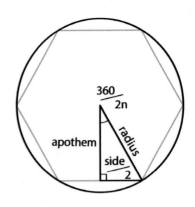

FIGURE 7-3: Angles and dimensions of an inscribed polygon.

$$\text{Area of this circle} = \pi\, r^2 = \pi\, (1)^2 = \pi.$$

We know now that we can see how far off we are based on how different the area of the polygons are from π.

AREA OF INSCRIBED POLYGONS

Earlier in the chapter, we saw that the inscribed polygon can be broken up into right triangles. We want to get the areas of each of these right triangles so that we can find the overall area of the polygons. If our polygon has *n* sides, each of which is broken up into two identical (mirror image) right triangles, each angle in the center of the circle will be 360°/2*n* (Figure 7-3). We also have to be given the radius of the circle our polygon is inscribed in, since everything scales with that.

Now we have one angle of a right triangle in addition to the right angle (which makes it easy to figure out the third angle), and its hypotenuse (the radius of our circle). We want to get the lengths of the sides of the triangles. As we saw in Chapter 6, we can do that with the sine and cosine of the known angle.

The *apothem* is the adjacent side of the angle marked in Figure 7-3 — the height of this triangle, in other words. The side opposite the marked angle is equal to base/2. We can use these two facts to get the apothem and the length of the side in terms of that angle and radius of the circle.

$$\cos(360°/2n) = \text{apothem/radius}$$

Multiplying everything times the radius, we get:

apothem (or height of the triangle) = radius * cos(360°/2n)

The base of the triangle (side/2) is the side of the triangle opposite the angle at the center of the circle, so we can use sine to figure it out.

sin(360°/2n) = (side/2)/radius

Again multiplying everything by the radius, we get

side/2 = radius * sin(360°/2n)

The area of each of the small inscribed triangles is (from Chapter 5, when we saw that the area of a triangle is 1/2 base* height):

Area   = 1/2 * base*height
          = 1/2 * (side/2) * apothem
          = 1/2 * radius * cos (360°/2n) * radius * sin(360°/2n)

We can make this a little simpler by realizing that 360°/2n = 180°/n.

Area of each triangle = 1/2 * radius^2 * cos(180°/n) *sin(180°/n)

There are 2*n* of these little triangles, so the area of the whole polygon is:

Area of the polygon   = 2*n* * 1/2 * radius^2 * cos(180°/n) *sin (180°/n)
                          = *n* * radius^2 * cos(180°/n) * sin(180°/n)

Back at the start of all this, we wanted to see how close these polygons were coming to the area of the circle they are inscribed in. To make things easy, let's say that is a circle of radius = 1. Therefore, its area would just be

Area of the circle = π * radius^2 = π * 1^2 = π

A circle of radius 1 is often called a *unit circle.* It comes up often in various branches of math since it is handy for comparisons like this. Here is a table of the areas of the inscribed polygons for a circle with radius =1, for the four polygons we showed in Figure 7-1. We can see that as the number of sides gets bigger, we are creeping up on the value of pi.

Number of sides, n	180°/n	Formula (for radius =1) $n * \cos(180°/n)\sin(180°/n)$	Inscribed polygon area
3	180°/3 = 60°	3 * cos (60°) * sin(60°)	1.299
4	180°/4 = 45°	4 * cos(45°) * sin(45°)	2.000
6	180°/6 = 30°	6 * cos(30°) * sin(30°)	2.598
12	180°/12 = 15°	12 * cos(15°) * sin(15°)	3.000

AREA OF CIRCUMSCRIBED POLYGONS

Let's try to develop the area of the circumscribed polygons. Now the radius of the circle is the height of the triangle. The base is half the side (Figure 7-4). As with the inscribed triangles, we know that the angle at the center of the circle of each of these right triangles is 360°/2n, which we can simplify to 180°/n.

Now we can proceed similarly to the way we got the area of each of the triangles making up the circumscribed polygon. For the circumscribed polygon, each triangle's area is:

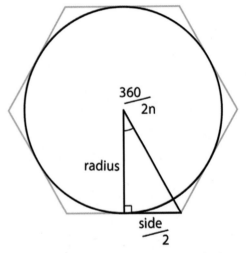

FIGURE 7-4: Dimensions of the circumscribed polygon.

$$
\begin{aligned}
\text{Area} \quad &= 1/2 * \text{base} * \text{height} \\
&= 1/2 \text{ radius} * \text{side}/2 \\
&= 1/2 \text{ radius} * \text{radius} * \tan(180°/n) \\
&= 1/2 * \text{radius}^2 * \tan(180°/n)
\end{aligned}
$$

The area of the whole polygon is 2n times the area of this triangle (2n because there are two right triangles for each side of the polygon). This table lays it out for a polygon circumscribed around a circle of radius = 1. The area of that circle, as we saw earlier, is just pi. As the polygon has more sides, we can see that the area of the polygon also approaches pi.

Number of sides, n	180°/n	Formula (for radius = 1) n * tan (180°/n)	Circumscribed polygon area
3	180°/3 = 60°	3 * tan (60°)	5.196
4	180°/4 = 45°	4 * tan(45°)	4.000
6	180°/6 = 30°	6 * tan(30°)	3.464
12	180°/12 = 15°	12 * tan(15°)	3.215

The inscribed and circumscribed areas are bracketing pi, with the inscribed areas always a bit smaller than the circle and the circumscribed areas a bit larger. As we can see in this table, we get to within a few percent of the area of the circle with 12 sided inscribed or circumscribed polygons.

Number of sides, n	Inscribed polygon area	Circumscribed polygon area	Area of circle
3	1.299	5.196	3.1416
4	2.000	4.000	3.1416
6	2.598	3.464	3.1416
12	3.000	3.215	3.1416

LEONARDO'S PUZZLE

Leonardo da Vinci and others developed a variant on the proof above. We know that the circumference of a circle of radius r is $2\pi r$. We want to prove that the area of a circle is πr^2. If we were to make a rectangle half the circumference long and as tall as the radius, can we show that the area of the circle and that rectangle are the same? This puzzle will show you how that works.

First, print out the model created by the OpenSCAD model **areaWedges.scad**. The model allows you to change the size of the model by making the radius value, **r**, bigger or smaller. We would recommend no smaller than its default of 30 mm. The number of sides is the variable **sides**, currently defaulted

FIGURE 7-5: Leonardo's puzzle, assembled

FIGURE 7-6: Leonardo's puzzle disassembled

FIGURE 7-7: Leonardo's puzzle rearranged into a rectangle

to 6. We will walk through how to use the 6 sided model. First, we print out a circle and its inscribed polygon cut into triangles (here, a hexagon). One of the triangles is split in half (Figure 7-5).

Next "unroll" the circle to get a long row of triangles. Then flip half of them upside-down and fit them into the lower half (Figure 7-6). The little tabs should line up to make that easier. Both of the half-triangles go upside-down. Note that you need to flip half of the pieces upside-down for the alignment tabs to work.

Now, you can make a rectangle that is π long by r high. The half-triangles help square off the ends. Figure 7-7 shows it assembled on a little L-shaped stand to help align it. (The curved parts of the bottom triangle are hidden by the stand.)

Measure the long side of the rectangle you get - is it indeed pi times the radius of the circle? Try it with more (or fewer) pieces by changing the variable **sides** in **areaWedges.scad**. Did you get closer to pi times the radius of the circle with more sides?

FIGURE 7-8: Finding the maximum distance across the can (the diameter)

FIGURE 7-9: Holding the maximum value to measure it

MEASURING PI WITH STRING

An easy alternative to find the value of pi is to use something round (like a soup can). Carefully measure the diameter of the can by stretching a string across the top (Figure 7-8). To do this, hold one side of the string tight on one side while holding the other side more loosely. Swing the looser side across the can, letting the string slip through your fingers as you go around the circumference and your hands get further apart. The farthest apart your two hands get is the diameter. Once you pass that point, the string will no longer be pulled farther through your fingers (Figure 7-9), and the length between your fingers will be the diameter. Measure it and write it down.

Then wrap a piece of string around the can. Measure the length of the string and divide the value you get by the diameter you measured. You should get close to pi, since the circumference around the can equals pi times the diameter.

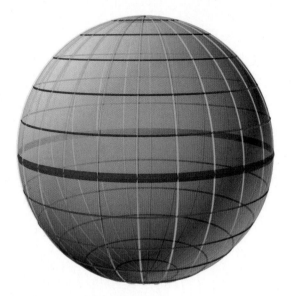

FIGURE 7-10: Latitude and longitude lines.

LATITUDE AND LONGITUDE

We've now learned enough math to figure out where we are on the earth's surface. In the next few sections of this chapter, we are going to build your ability to use observations to be able to estimate where you are on earth based on the positions of the sun and other objects in the sky, plus a few common tools like a protractor and ruler.

About 2100 years ago, the Greek astronomer Hipparchus is credited with figuring out the basics of measuring what we now call latitude and longitude. He is also credited with early work on what became trigonometry. The 1989 European Space Agency's Hipparcos astrometry mission was named in his honor (astrometry being the precise measurement of the locations of the stars in the sky) as well as craters on the moon and Mars.

On a globe, longitude lines (called meridians) run from pole to pole, marking distance east and west. These are the green lines on Figure 7-10. You could start counting anywhere since the earth is round, but by convention a line running north and south through Greenwich, England, is considered to be the zero meridian. Distances to the west of Greenwich are treated as negative numbers, or "Longitude West". Los Angeles, California is about 118° west, or -118°. 118° east runs through the middle of Nanjing, China.

The red lines (circles) on Figure 7-10 are latitude, marking distance north or south of the equator (the heavy red line) which is 0°. The poles are plus (north) and minus (south) 90° latitude. Typically north is at the top of this imaginary system, and south at the bottom.

Let's walk through figuring out your latitude and longitude just from simple observations of the sun at local noon. We will take it one step at a time, using the information about sine and cosine we just learned in Chapter 6. Of course, you can look up your address in an online map program, but what's the fun of that? You can check your answer that way, though.

FINDING YOUR LATITUDE

The two numbers you need to figure out your latitude are the *solar declination* and the *elevation angle*. Elevation angle is the angle of the sun above the horizon at its highest point for the day. The solar declination is the angle between a line connecting the centers of the earth and sun, and the earth's equator, and it adds or subtracts as much as 23.44° to the elevation angle over the course of the year. (We discuss how to compute it for a particular day of the year in a later section in this chapter.) Both of these vary over the year (and longer). Let's see how to get those numbers in a sunny south-facing window, or your backyard.

Your latitude measures how far north or south you are from the equator. It is measured as an angle between two imaginary lines: one that goes through the center of the earth at the equator, and one from the center of the earth to your location (marked by the red dot in Figure 7-11).

We talk about being located at so many degrees north or south latitude. Los Angeles, for example, is at about 34° north. Latitudes south of the equator are negative numbers; the South Pole is at 90° south, or -90°. Most online maps have a way to show latitude and longitude. Google Maps will show latitude and longitude when you click on a particular spot.

As it turns out, the earth's north pole doesn't poke straight up at right angles to the earth's orbit. Instead, it is tilted at about 23.44° to that right angle. If you imagine the solar system with the sun in the middle, the earth's axis always points off to one point in space all year long.

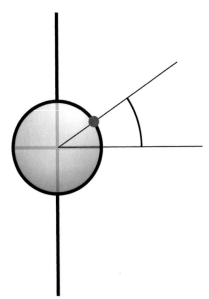

FIGURE 7-11: Latitude of the point denoted by the red spot (north pole shown pointing directly upwards).

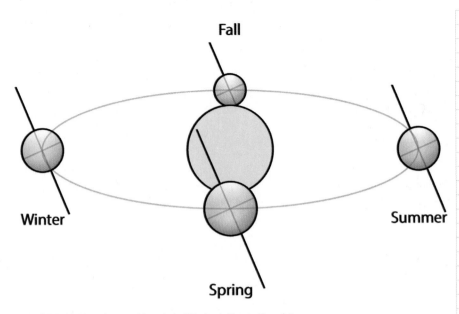

Fall

Winter

Summer

Spring

FIGURE 7-12: How the earth's axis is tilted relative to its orbit.

Figure 7-12 shows how that works, noting the seasons for the northern hemisphere.

FIGURE 7-13: A modeling clay earth with axis and northern hemisphere latitude noted.

We will talk through this in diagrams here, but you might want to make yourself a ball out of something like PlayDoh or modeling clay and stick a toothpick through it (Figure 7-13). The toothpick is the earth's axis. Draw a line around the "equator" of your modeling clay earth. This line should be around the center, and shouldn't appear to wobble when you twist the toothpick between your fingers. Then, make a little mark on the earth to show where you are, or use a straight sewing pin poked in where you are down toward the center of the earth. The angle of the pin can help you think about your latitude.

Put a lamp on a table in the middle of a room so you can walk around it. Then go ahead and circle it once (you're the earth orbiting the sun). Keep the toothpick pointed to one corner of the room. Think about what happens at the equator, and where you are, at local noon on the first days of fall, winter, and spring.

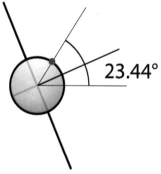
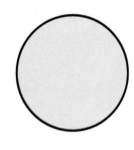

FIGURE 7-14: The sun and the earth alignment on the winter solstice (day 1 of the solar year).

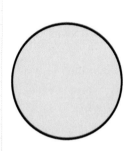
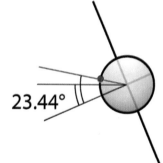

FIGURE 7-15: The sun and the earth alignment on the summer solstice (day 183 of the solar year).

Let's go into a bit more detail for the northern hemisphere example. On the shortest day of the year (the first day of winter, which is the winter solstice) the sun will appear to be 90° minus your latitude minus 23.44° off the horizon. In Los Angeles, that would be 90° - 34.1° - 23.44° = 32.44°. So on December 21st the sun's highest point is only about a third of the way up the sky, even in sunny Los Angeles. The winter solstice is often used as the first "day of the year" when we calculate sun elevation angles, since it is the lowest "high" point of the sun (Figure 7-14).

We call the longest day of the year the first day of summer, or the summer solstice. If you look to the south at noon on that day, the sun will appear to be 90° overhead, minus the difference between your latitude and 23.44°, as we can see in Figure 7-15. In Los Angeles, this would be 90° - (34.1° - 23.44°) = 79.34°, as shown in Figure 7-15. This usually happens on June 21, about 182 days after the solstice (183 in leap years, but it's really about 182 plus a quarter day).

On the first day of spring or fall, the earth's tilt is not toward or away from the sun, and the highest point of the sun in the sky is just 90° minus your latitude.

There is a formula you can use to figure out the maximum elevation of the sun in-between these four times of the year:

$$\text{Max elevation} = 90° - \text{latitude} + \text{solar declination}$$

That means that to find latitude, we need to find the sun's maximum elevation above the horizon for the day. We also need to calculate the solar declination for that day, which depends on how many days have passed since the last winter solstice. We'll show how to calculate that in a later section in this chapter. We can find the sun's elevation by measuring it, as we'll show in the next section.

Incidentally, the earth's axial tilt varies slowly over time, and is currently decreasing. Over many millions of years, it has varied from about 22° to 24.5°. Unless you are doing precision navigation, you can safely ignore this variation over a human lifetime. However, astronomers often think about the long term.

ELEVATION OF THE SUN AT LOCAL NOON

Let's measure the *elevation* of the sun, the angle between it and the horizon when the sun is at its maximum height for the day. First we need something that will cast a shadow that is easy to measure. The fancy name for this is a *gnomon*. A toothpick in modeling clay or something else pointy held vertically will work, or we have a 3D printable one that will be a little easier to use. If you are using our 3D printed gnomon, first create it from the model in **gnomon.scad**

Measuring elevation is not as simple as going outside when the clock says noon. Depending on where you are in the time zone, the sun might be at its highest point as much as an hour before or after noon on the clock. In some extreme cases, like the far west of Alaska, it might be even more if a state is kept in one time zone for convenience.

For that reason, you should start doing the following measurements shortly after 11 AM if you are currently observing standard time, and at about noon if you are someplace using daylight savings (summer) time. You might have to do this for as much as two hours. If you are using daylight savings (summer) time, start measuring at noon on your clock and you might have to go as far as 2 PM Daylight Savings time.

FIGURE 7-16: 3D printed gnomon

If you know you live near a time zone boundary, you might be able to think through how that works out and make the following process shorter. If you want to assume nothing, however, here is what you should do. The following directions are for standard time; shift them all an hour later for daylight savings time.

Take a piece of paper and your 3D printed or toothpick-and-ruler gnomon outside. Every five minutes, starting at 11:00, do the following:

- Put the gnomon on a sturdy (and level) surface that will be in the sun for at least two hours.
  - Arrange it so that the shadow of the vertical (shorter) part falls on the horizontal part. Read off the solar angle directly (Figure 7-16).
- If you are using a toothpick in clay, first place the toothpick right over the zero marking on the ruler. Then pack around it with a bit of modeling clay to make it stay upright. Be sure that the toothpick is vertical and making a right angle with the ruler, as shown in Figure 7-17. You want to use something pointy so that the shadow of the tip is crisp and easy to read. If you can figure out how to stand it up, a pencil with its tip up would work too.
- Measure the shadow. You may need to rotate the gnomon to have the shadow fall squarely on it. Be sure you are measuring from the center of the toothpick or pencil, not from the edge of the clay!
- Re-do this measurement every five minutes until the shadow reaches a minimum height and starts getting longer again. (If it is already getting longer when you start, try again the next day, but start earlier.)

FIGURE 7-17: Using a ruler, toothpick, and clay as a gnomon

- Write down as accurately as you can the time when you saw the shortest shadow. You'll need that time to calculate your longitude later in this chapter.

In Chapter 6, we learned about sine, cosine, and tangent. The tangent (opposite side divided by adjacent side) of the angle of the sun above the horizon is the height of the vertical part of the gnomon (the toothpick) divided by the length of the shadow. For example, if the gnomon is a 6.5cm tall toothpick stuck into a bit of modeling clay so it stands on its own (rather than being stuck in the ground), and the shadow is 6.8cm long, the angle whose tangent is 6.5/6.8 = 0.96 is 43°.

Once we get this angle, we have one of the key numbers we need for our calculation of latitude: the maximum elevation angle of the sun today. Write down the angle you got from your observations. Remember that we are using this equation:

Max elevation = 90° - latitude + solar declination

Now we just need solar declination to calculate latitude. For our example we got 43° for solar elevation, and we would have

43° = 90° - latitude + solar declination.

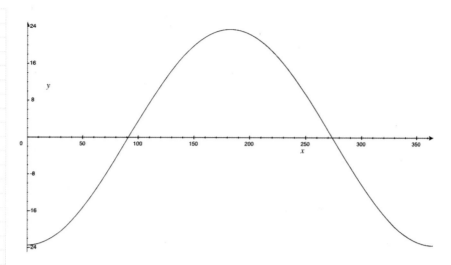

FIGURE 7-18: Solar declination in degrees versus days since the winter solstice

SOLAR DECLINATION

The solar declination is the angle between a line connecting the centers of the earth and sun, and the earth's equator. It adds or subtracts as much as 23.44° to the elevation angle over the course of the year. To do it right, we would need to use a complicated formula that takes account of the earth's orbit not being perfectly circular, in addition to the tilt of the earth's axis. Fortunately, though, we can get within a degree or two with this approximation:

> Solar declination (in degrees) = - 23.44° * cos ((360°/365) * (days since the first day of winter)).

There are 360° in a circle, and it takes 365 days to go around the sun in a year. That's where the 360°/365 comes in since that is how many degrees the earth goes around its orbit per day. We multiply that times how many days it has been since the sun was at its most extreme low point, when it is 23.44° below where it would be if the earth's axis pointed straight up and down. This day is called the *winter solstice* in the northern hemisphere, and we also think of it as the first day of winter. In other words,

> (360/365) * days since the winter solstice (which is 10 + days since Jan 1)

equals the fraction of the almost-circle of earth's orbit we have gone around since the first day of northern winter. If we plot declination versus days since the first day of winter, we get the graph in Figure 7-18.

Solar Declination Example

Let's say we want to know the solar declination on October 22, 2023. That turns out to be day 295 in 2023, which is not a leap year. So the solar declination on October 22, 2023, is:

Solar declination (in degrees) =
$-23.44° * \cos((360/365) *(295+10))$
$\qquad = -23.44° * \cos(300.8°)$
$\qquad = -23.44° * 0.51237 = -12.0°$.

Since we need to subtract this number of degrees off the altitude of the sun on the first day of winter (our starting point), it is a negative number.

Therefore the maximum sun elevation on October 22, 2023 anywhere in the world will equal: $90°$ - latitude - $12°$, or $78°$ - latitude.

Notice that this means at latitudes north of $78°$, the sun will not be above the horizon on October 21 at all. This line of latitude runs roughly through the middle of Greenland. You can play with this and a globe or online maps to see what part of the world will start missing the sun altogether on a particular day.

On the first day of winter, we subtract off $23.44°$ from the sun's highest point above the horizon. About 182 days later, we add $23.44°$. It's tricky to think about this in the Southern Hemisphere, where the negative latitude gives us sun angles of more than $90°$. This means the sun will be in the north during most or all of the day.

Suppose we wanted to be a little more accurate and take account of the fact that the earth's orbit isn't quite 365 days. It is more like 365 and a quarter days, which is why we have leap years every four years.

Also, the winter solstice isn't exactly at local noon in any given place. In 2022, it is at 1:48 PM Pacific Standard Time. To correct for that, we can use fractional days. If we are measuring around noon Pacific Standard Time (close to 1 PM Pacific Daylight Time, in October 2023, when local noon is), that's about 1 hour and 48 minutes less than a full day from than the solstice was, or 0.075 of a day less. We could use both these corrections to get:

Solar declination (in degrees) =
$-23.44° * \cos((360 / 365.25) * (295 - 0.075 + 10)) = -11.9°$

This difference is probably less than the accuracy we will get in measuring our elevation angle, but good to know that our approximation is close. We will call it $-12°$ in our example.

To get days since the first day of winter, remember that winter starts on December 21, more or less. There are 10 days left till the end of that year. Then, you can look up a "day of year" calendar online (for example, at **timeanddate.com**, under "other options" for calendars), which lists the days from January 1 (which is day 1) to December 31 (day 365 or 366, depending on whether it is a leap year). Be sure you find a calendar for your current year so you get the leap year correction.

If you wanted to be a little more accurate, the earth's orbit around the sun is about 365 and a quarter days long so you could use 365.25, but we'll keep it simple here. Also, the winter solstice is an event in time that is not right at noon. You could look up when the last winter solstice was and calculate in fractions of a day. We'll do that in the example we calculate in the next section.

Calculate the solar declination for the current day of year and write it down. Now we have everything we need to find our latitude, since we are working with this formula:

Max solar elevation = 90° - latitude + solar declination

CALCULATING LATITUDE

Now let's put that together for our example near Los Angeles on October 22, 2023, the 295th day of the year. We measured that the maximum sun angle is 43° and we calculated solar declination = -12°. The equation for latitude is:

Maximum sun altitude = 90° - latitude + solar declination, so in our Los Angeles in October example:

$$43° = 90° - \text{latitude} - 12°$$

If we add the latitude to both sides, we get

$$43° + \text{latitude} = 90° - 12°$$

Now subtract 43° from both sides

Latitude = 90° - 12° - 43° = 35°, which is very close to the actual Los Angeles value of 34.1°.

ARCTIC AND TROPICS

We can also now figure out how far north you need to go to have the sun never rise above the horizon in the winter. We can just set the maximum sun altitude to zero like this:

Maximum sun altitude = 90° - latitude - 23.44° (on the first day of winter).
If this = 0, 0 = 90° - latitude - 23.44°
Or latitude = 90° - 23.44° = 66.6°

At latitudes higher than that, the sun does not manage to get above the horizon at its highest point of the day on the day that has the least number of daylight hours. That latitude is called the Arctic Circle. It is a little north of the city of Fairbanks, Alaska, and runs through Norway, Sweden, Finland and Russia. The farther north you go, the more days of the year the sun does not rise at all.

The same line of latitude exists in the southern hemisphere. It is called the Antarctic Circle and it exists at 66.6° south (or -66.6°) neatly circling the continent of Antarctica. The southernmost of the South Shetlands Islands of Antarctica, for example, are close to the Antarctic Circle.

When you get to the north or south pole, it is dark for six months at a time. Take a globe and figure out why. Try to work out these special cases (answers are at the end of the chapter).
- What is the maximum sun altitude at the South Pole (latitude -90°) on the first day of southern summer (first day of winter in the north, with solar declination = -23.44°)? What direction (north, south, etc) is the sun?
- Latitude 23.44° north is sometimes called the Tropic of Cancer, and the equivalent in the south (at 23.44° south) is called the Tropic of Capricorn. On the first day of summer in the northern hemisphere, what is the solar elevation at local noon for a place on the Tropic of Cancer?

FINDING YOUR LONGITUDE

If we want to know where we are on earth, we need to know both our latitude north and south of the equator, and our longitude east and west. However, calculating longitude is tricker, and can't be done entirely from what you can observe with rulers and protractors. There is no way to tell the difference just from the sun or stars whether it is an earlier time of day, or whether you are just farther west. To find longitude, you need some known point of reference, and an accurate, portable way to keep track of time so you can compare the height of the sun (or a star) relative to where it would be at that arbitrary other reference place. We will use the time of day when the sun is highest as our reference point, just as we did to find latitude.

The north-south reference longitude line is called the *Prime Meridian*, and is defined to be 0° longitude. It had been proposed at various times to run through Rhodes by the Greeks and Paris by the French, and for years different areas kept their own time (and therefore longitude) reference lines. In a nod to the British Empire's dominance of the seas, in 1844 an international agreement set the current location of the Prime Meridian, which runs through the grounds of the former Royal Observatory (now a museum) in Greenwich, England.

HISTORICAL EFFORTS TO FIND LONGITUDE

Our friend Hipparchus got tricky and proposed using observations of eclipses, which occur at known times, to do some rudimentary longitude measurements. To do that, observers in two places would see how the eclipse went past them, and try to figure out some way of comparing what they saw. It's not clear if he figured out how to make that work. Columbus is reported to have tried it too, although it is controversial whether he did. Not a lot happened for a long time after Hipparchus, since clocks good enough to tell time for navigational purposes had to wait about 1800 years.

At the height of Britain's naval power in the 1700s, there was a big prize (the Longitude Prize) for the first person to come up with a good way to measure time on a ship, and thus longitude. A person who was more or less a garage inventor at the time, John Harrison, eventually solved the problem of keeping accurate time on a rolling ship in all weather. He came up with clever inventions including bimetallic strips, but because of politics and intrigue, never won the prize (nor did anyone else) although Parliament eventually supported his work.

APPROXIMATING LONGITUDE AND THE EQUATION OF TIME

Unlike British sailors in the 1700s, you have accurate clocks at your disposal (by historical standards), so you would think you can get your longitude quite simply by seeing when the sun is at its highest point for the day and comparing that to noon in Greenwich, when presumably the sun would also be at its highest point that day. However, there are complicating factors that are captured by the *equation of time*, which is the difference between the time when a sundial would measure local noon versus noon on an accurate clock that measures the average time of local noon over the course of a year.

Your longitude in terms of hours and minutes (we'll convert it to the more commonly-used degrees later) is the time of highest sun elevation in Greenwich, England minus your local time when the sun is highest (on the same day).

Now, there's a small problem here. You don't actually know when the sun was highest in Greenwich, since the observatory there is unlikely to go out and look for you. Therefore, you need to adjust it by the value of the equation of time for that day of year.

If the earth's orbit was a perfect circle (which it isn't) you could just subtract the time the sun was highest for you from the equivalent event in Greenwich, and you would have your longitude. Alas, it's not that simple. Like solar declination for latitude, the equation of time varies periodically over the course of a year. We'll talk more about where the equation of time comes from in Chapter 12, but for now let's just look it up, by estimating from the graph in the Wikipedia article "Equation of time" or searching on "equation of time calculator." Here is what we need to find longitude.

> Your longitude =
> - Noon in Greenwich (on a clock)
> - Minus the equation of time for the day of year you are doing your measurement
> - Minus how many hours different your time zone is from Greenwich
> - Minus your local time when the sun was highest

The first two items, added together, give you the time that the sun is highest in Greenwich on that day of the year. The last two items added together are your local time the sun is highest, converted to Greenwich time.

LONGITUDE CALCULATION EXAMPLE

It's a little hard to keep the pluses and minuses straight, so let's do an example you can follow along with before you do your own. Let's say we measured the sun at its highest point at 11:35 AM United States Pacific Standard Time on October 22 (305 days since the winter solstice, or day of year 295; look carefully at the equation of time chart you use to see which convention they are using).

First we need to figure out how many time zones we are from Greenwich, England. Pacific Standard Time is 8 hours earlier, or west, of Greenwich Mean Time (GMT), which is the standard time at 0° longitude. Part of figuring out longitude is figuring out how far we are from the imaginary line marking the average location of the time zone, too. Time zone boundaries zig and zag to avoid having parts of states and countries in different time zones, but the sun doesn't care about that. (Really, you are just trying to convert your time to Greenwich time.)

Near Los Angeles, California, we observed that the sun was highest at 11:35 AM on October 22.

If we look up the equation of time for October 22, it's at just about its largest (+16 minutes) value on October 21. In Greenwich, which is by definition right on the time zone line, we would subtract 16 minutes from noon to get the time of maximum sun height, or 11:44.

So our longitude =
    12:00 (noon in Greenwich)
    Minus 00:16 (16 minutes positive equation of time on Oct. 21)
    Minus 8:00 (8 hours west)
    Minus 11:35 (local time of highest sun elevation)
    = -07:51 hours difference between California and Greenwich time

Or, to put it another way:

Longitude = 11:44 (time of highest solar elevation in Greenwich) - 19:35 (our local highest sun time, converted to Greenwich Time). If we subtract 11:44 - 19:35 we get that on October 22, the sun was at its highest point 07:51 later in California than in Greenwich.

Back in Chapter 6's "Degrees, Radians, and Pi" section, we saw how to convert hours, minutes, and seconds to degrees and fractions of a degree. Our -07:51 turns out to be -117.75°, which is correct to within our measurement error if we were only finding the time of highest sun elevation to within 5 minutes or so (equivalent to plus or minus 1.25° of longitude). We took our measurements in the Los Angeles suburb of Pasadena, California, which is at a longitude of -118.1°. By convention, longitude west is negative, and this might also be written 118.1°W.

If you are more than 12 times zones west of Greenwich, the convention is to start counting (at Greenwich) to the east, so we talk about 120°E longitude, not 240°W. However, to follow the instructions here, figure it out going east and add 360°. For example, if you followed these steps and got 242°W, that's the same as 118°E.

SUMMARY AND LEARNING MORE

In this chapter, we applied concepts from previous chapters to think about the area of a circle. We spent some time thinking about the biggest polygons that could fit into a circle, and the smallest ones that could fit around a circle, and used those to estimate pi and understand where the formula for the area of a circle came from.

This discussion of circles, plus our trigonometry basics in Chapter 6, gave us the background we needed to be able to figure out our latitude and longitude based on a measurement one can do with a toothpick, a ruler, and a sunny lunchtime. We will build on our latitude and longitude measurements in Chapter 12 with more astronomy explorations. If you want to learn more about why measuring longitude was so challenging, and the race to build clocks to fix the problem, check out Dava Sobel's book, *Longitude* (1995, Walker Publishing).

Meanwhile, next up will be going into the third dimension. Chapter 8 will explore the volumes of various solids, and how to use the ideas of volume and density to see how Archimedes kept his king from being cheated.

ANSWERS

Here are the answers for the activities in this chapter that we don't solve in the text of the section.

ARCTIC AND TROPICS

What is the maximum sun altitude at the South Pole(latitude -90°) on the first day of southern summer (first day of winter in the north, with solar declination = -23.44°)? Where does that mean the sun is in the sky?

Answer: Maximum sun altitude = 90° - (-90°) - 23.44° = 180° - 23.44° = 156.56°. That number is hard to think about, so instead think of 180° as going all the way from pointing due south, to pointing due north. We are going 23.44° less than that, so the sun would be 23.44° above due *north* at its highest point on the southern summer solstice. To see what would happen the rest of the day, check out the "Midnight Sun" article in Wikipedia. Latitude 23.44° north is sometimes called the Tropic of Cancer, and the equivalent in the south (at 23.44° south) is called the Tropic of Capricorn. On the summer solstice in the northern hemisphere, what is the solar elevation at local noon for a place on the Tropic of Cancer?

Answer: The sun would be directly overhead (at an elevation of 90°). The definition of these latitudes is that it is the farthest north or south you can go from the equator and have the sun be directly overhead on the first day of local summer.

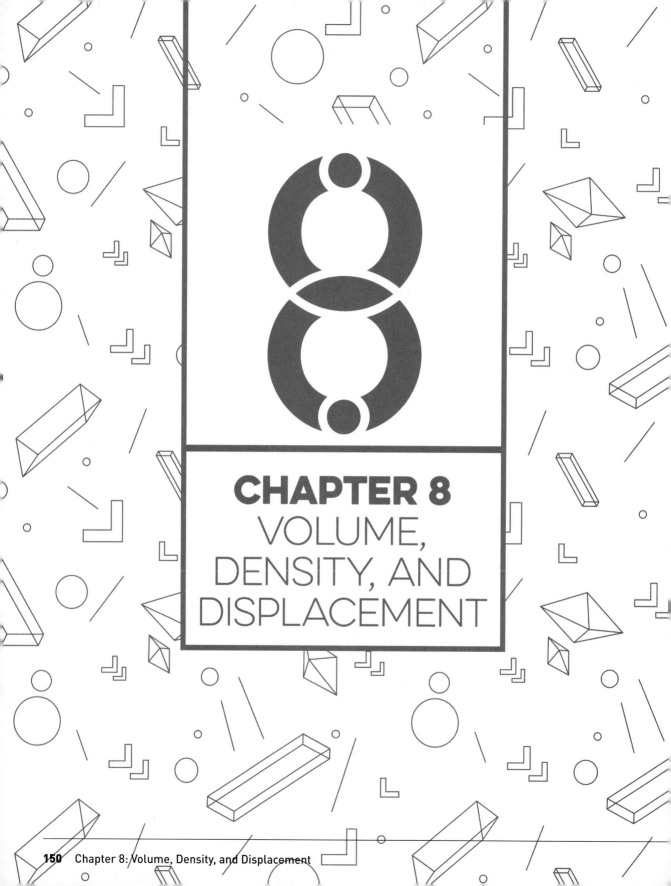

CHAPTER 8
VOLUME, DENSITY, AND DISPLACEMENT

In this chapter, we will explore three related topics: volume, density, and displacement. First, we calculate *volume*, which tells us how much space there is inside an object. Then we'll move on to *density* (how much mass there is in a given volume). We will wind up showing you how to measure *displacement*, how much water is equivalent in mass and/or volume to a solid object. (There is a concept of displacement in more advanced geometry, too— it is the shortest distance between two locations of a moving object. But we will not be using the word in that sense here. We are sticking to the more common nautical definition.)

You might wonder why we are doing what you might think of as physics in a geometry book. We want you to experience how geometry is used every day. One thing that is particularly fun about the content in this chapter is that we will be able to show you how you can calculate and measure volume, density, and displacement in the real world, and see that your calculations work beyond just having them on the page.

Like a lot of geometry, ways to calculate the volume of various 3D shapes have been known for well over 2000 years, since there are so many practical everyday applications for knowing how much a box or jar can hold. You also might want to know the contents of a container without necessarily opening it, so how much a certain volume of grain or gold should weigh is handy to know, too.

Ancient Greeks or Egyptians could tell if two masses were equal (or heavier or lighter relative to each other) by using a *balance*. A balance is a simple scale with two pans on a bar, connected to a pivot; if the two pans contain the same mass, the bar connecting them is level. If one side of the balance is heavier than the other, the crossbar rotates on the pivot, and that side sits lower. A balance is the instrument we see statues of Justice holding. But a balance couldn't do very good absolute measurements. That is, they couldn't easily tell that something had a mass of precisely 8.2 grams.

To measure mass and densities within those limitations, a little over 2000 years ago the Greek mathematician and engineer Archimedes came up with the concept of *displacement*. Displacement lets us use the known density of water versus other materials to see if golden crowns, for example, really were what the seller claimed them to be.

Archimedes' discovery is one of the classic science-discovery stories. It might or might not be a true story, since it has come down to us indirectly from sources written long after the fact. But, it's such a fun story that it *should* be true, so, with that grain of salt...

Archimedes lived in the city of Syracuse, on the island of Sicily. At the time Sicily was part of Greece. He was pretty oblivious to politics and a lot of other daily-life trivialities, but was so good at what he did that whoever was in charge at the time wanted his advice. For example, King Hiero II of Syracuse suspected that his crown-maker had kept some of the gold that was supposed to be for a new crown, replacing it with a less valuable metal. The king turned to Archimedes for an assessment.

3D Printable Models Used in this Chapter

See Chapter 2 for directions on where and how to download these models.

sphere_cone_volume.scad
Prints out a cone, half-sphere, and cylinder with volume ratio 1:2:3

prism.scad
Prints an n-sided prism (flat top and bottom) with a specified volume

pyramid.scad
Prints an n-sided pyramid (becomes a cone for n >100) with a specified volume. The top can also skew away from the center while maintaining the volume.

3_pyramid_puzzle.scad
Prints 3 pyramids which collectively form a cube

Other supplies for this chapter
- A scale (ideally, a postal or kitchen scale that can measure to fractions of a gram)
- About 1/4 cup of table salt, sand, or other fine granular substance that does not pack down when compressed.
- A calculator
- Measuring cup or graduated cylinder with detailed markings, ideally in mL
- A basin of water
- A few metal coins and other objects that can be submerged in water

Archimedes thought about it, and he knew that gold was denser than any other metal they might have used. If he knew the volume of the crown, he could compare it to an equal mass of gold. Since the crown was a complicated shape, though, there was no easy way to measure its volume. Then, one day, getting into a bathtub, he saw that some water sloshed out. He realized that he could put this complicated-shaped crown in some water, and find out its volume because an equal volume of water would be displaced. As we'll see in this chapter, this allows us to get on the path to measuring the density of an object of unknown composition.

In the most common version of the story, he was so excited about this that it is said that he ran naked through the streets of Syracuse, shouting, "Eureka!" (Greek for "I have found it"). And so today we have "a eureka moment" as a moment when a great discovery comes about. Your authors live in California, whose state motto is, "Eureka!" (harkening back to our gold rush days).

The crown maker? The tale goes that he was indeed crooked, and presumably dealt with. Not a eureka moment for him.

Hopefully, we can lead you to your own eureka moments in this chapter, although we suggest dressing first before running off to tell your friends. Let's see how much we can learn about volume, density, and displacement with very simple measuring devices, and also see how Archimedes solved the crime.

VOLUME

There are a lot of practical applications of knowing the volume of 3D objects (specifically, cones, spheres, and cylinders). Back in Chapter 3, we learned about regular polyhedrons and how to make them in OpenSCAD.

Not surprisingly, we measure volume in units of length cubed — that is, extended into all three spatial dimensions. A standard unit of volume is the liter, which is precisely one-tenth of a meter squared. A thousandth of a liter, a milliliter (mL) is also equal to 1 cm^3.

It turns out that there are simple relationships among the volumes of a sphere, cylinder, and cone. Let's look at each in turn, and then tie them together.

VOLUME OF A RECTANGULAR SOLID

If the sides of a polyhedron are rectangles instead of squares (for example, a box), we can pick one side and call it the base. Then the volume of this *rectangular solid* is just the area of the base times the height, or length * width * height.

> Volume = base * height

A *cube* is a six-sided regular polyhedron. All its sides are squares. In this case, since the length, width, and height are all the same, we can use any side and just cube it.

> Volume = side^3

In OpenSCAD, all rectangular solids are created with the "cube" function. For example, a rectangular solid that is 10 by 30 by 20 mm would be created in OpenSCAD with this line:

```
cube([10, 30, 20]);
```

An actual cube could either be the same line with all the numbers the same, or just this (for a cube 20mm on a side):

```
cube(20);
```

Figure 8-1 shows the results of these two single-line models. We haven't

FIGURE 8-1: A rectangular solid and a cube.

provided a model for rectangular solids given that it is probably simpler to type in one line than to hunt around in the repository for them.

VOLUME OF PRISMS

Now we'll tie together a few ideas that will let you see that these formulas hold for some other shapes as well. Pieces of this analysis go back 2300 years or so to Archimedes in ancient Greece. The form we will talk about here though is attributed to Bonaventura Cavalieri, who was active in the mid-1600s, just before Newton. Some of the ideas in this chapter are a good lead-in to calculus as well.

Imagine that we have a *prism*. A prism is a 3D shape with identical, parallel polygons at both ends, and flat faces connecting the sides. For example, a triangular prism has identical triangles on top and bottom, and three flat rectangular sides. A cube is a prism, too, which happens to have all its sides the same.

One of these sides is considered the base, and the height is measured perpendicular to the base. As with a rectangle, the volume of a prism is just the area of the base times the height. Therefore, if two prisms have the same base area, regardless of the shape, and the same height, they will have the same volume.

If the top is directly over the base (and thus the sides make right angles with the base and top) it is called a right triangular prism (if the cross-section is triangular), a right rectangular prism (if the base is a rectangle), and so on.

FIGURE 8-2: Triangular right prism and oblique prism

FIGURE 8-3: Two stacks of coins, each with the same volume

The volume of an *oblique* prism is just the base times the height too. The top polygon of an oblique polygon is not directly above the base. However, the top and the base, as well as every horizontal cross-section in between, still have the same area. The vertical distance between the top and the base is still the same, therefore the volume remains the same.

Figure 8-2 shows a right triangular prism and an oblique triangular prism. These were both made with the model **prism.scad**. This model by default prints hollow shapes whose inner volume is the specified one. If you set the wall thickness (**wall** variable) to zero, it will instead create a solid object with the specified volume.

The volume stays the same even if the sides are not straight, as long as the edges all had the same curve, so that the cross-sections all remained the same. To demonstrate this another way, imagine a stack of identical coins. If we push some of them so that they aren't a straight cylinder, the volume of the stack (the sum of the volumes of all the coins) is still the same (Figure 8.3).

This is called *Cavalieri's Principle*, named for Bonaventura Cavalieri, a

17th-century Italian mathematician. The Principle can be stated like this: If you take a series of slices through two objects parallel to the base, you can compare the area of the cross-section where they were cut. If, for every possible parallel cut, the two objects have equal-area cross-sections, then the volumes of those objects are also equal, even if the shapes of the cross-sections are different.

MODIFYING THE MODELS

The model prism.scad and pyramid.scad have these variables:

- `v = 50000;`
  - volume in cubic mm (cc * 1000)
- `h = 30;`
  - height in mm
- `n = 6;`
  - number of sides (not including top/bottom); use 300 to get a cylinder
- `wall = 1;`
  - Wall thickness, in mm; should be about twice the line width for your printer. If **wall = 0**, the model will output the solid model. If **wall** is greater than zero, the pieces print out hollow and open, with an interior volume equal to the parameter **v.**
- `lid = false;`
  - Changing this parameter to "true" creates an additional removable lid that fits over the open end of the part. Note that these lids tend to add quite a bit to the print time of the part. (Default is no lid.) This variable is ignored if **wall = 0**.
- **offset** = [0, 0]; (right prism or pyramid), or
- **offset** = [30, 0]; (oblique prism shown here - any nonzero values work)
  - The x, y coordinate offset of the top of the prism relative to the bottom

In addition, the pyramid.scad model allows you to add a little ball on the top to make it easier to handle. The **ball** parameter specifies the radius. If **ball** is left set to its default value of zero, the point will still be rounded-off with a radius equal to the wall value.

Create a triangular (**n = 3**), rectangular (**n = 4**) and hexagonal (**n = 6**) prism, all with the same volume. Make one of them oblique. Try the exercise of filling them with salt one after each other to see if they are the same as

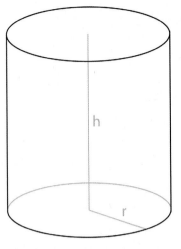

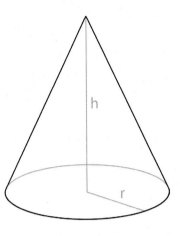

FIGURE 8-4: Anatomy of a cylinder   **FIGURE 8-5:** Anatomy of a cone

each other (as they should be). If they aren't, check that you didn't accidentally scale one of them in the slicing software.

VOLUME OF A CYLINDER

A cylinder is an object that has a constant circular cross-section, like a soup can. The sides are straight and parallel to each other, at right angles to the circular cross-section (Figure 8-4). There are skewed cylinders with sides that are not at right angles, and they can be treated just as we did oblique prisms. To get the volume of a cylinder, we multiply that cross-section times the height. Since the cross-section is circular, its area is πr^2. If we call the height h, then

$$\text{Volume of a cylinder} = \text{base area} * \text{height} = \pi r^2 h$$

VOLUME OF A CONE

A cone, like a cylinder, has a circular cross-section. However, this cross-section steadily gets smaller as we go from the base of the cone up to its point (Figure 8-5). Think of an ice-cream cone or a party hat. You might expect that the area of the base, πr^2 would be involved somehow. As it turns out, if a cone has height h, then

$$\text{Volume of a cone} = \tfrac{1}{3} * \text{area of the base} * \text{height} = \tfrac{1}{3}\,\pi r^2\,h$$

Notice that this is exactly ⅓ the volume of a cylinder of the same radius and height. The philosopher Democritus of Greece, who lived about 2400 years ago, is credited with being the first to realize this, although why it was true had to wait about 2000 years for Cavalieri. Archimedes (about 100 years after Democritus) expanded his ideas to get the relationships among the volumes of spheres, cones, and cylinders.

VOLUME OF A SPHERE

You can think of a *sphere* as the 3D equivalent of a circle. A ping-pong ball is a sphere. The Earth is more or less spherical. Technically, a sphere is defined as the set of points that are all the same distance

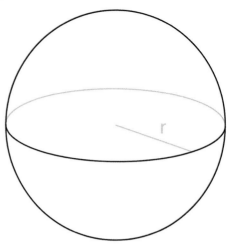

FIGURE 8-6: Anatomy of a sphere

(the radius) from its center (Figure 8-6). Given that, it's not too surprising that all you need to get the volume of a sphere is the radius, which we will call r.

Volume of a sphere = $\frac{4}{3}\pi r^3$

For example, if we had a sphere that had a radius of 2 cm, the volume would be

$$\frac{4}{3}\pi r^3 = \frac{4}{3}\pi * (2cm)^3 = \frac{4}{3}\pi * 8cm^3 = 33.51cm^3$$

If you wanted to measure all the way around a sphere (making a circle like the equator on Earth), that line is a 2D circle of radius r; therefore the circumference will be $2\pi r$.

VOLUME SPECIAL RATIOS

Let's review the volume formulas. Remember that r is the radius of the shape in question, and h is its height.

Shape	Volume
Sphere	$\frac{4}{3}\pi r^3$
Cylinder	$\pi r^2 h$
Cone	$\frac{1}{3}\pi r^2 h$

Now, we will explore a special case, when the height, *h* of the cylinder and cone are equal to the radius, *r*. What do the volume formulas become if we make *h* = *r*?

	Volume if height (h) = radius (r)
Sphere	$\frac{4}{3}\pi r^3$
Cylinder	$\pi r^2 r = \pi r^3$
Cone	$\frac{1}{3}\pi r^2 r = \frac{1}{3}\pi r^3$

We've created a model, **sphere_cone_volume.scad,** that creates a half-sphere, cylinder, and cone for this special case, where all three volumes are a multiple of $\frac{1}{3}\pi r^3$. Since the cone is the smallest, let's think about all the volumes as multiples of the volume of a cone.

	Volume compared to cone
Half Sphere	2 (full sphere would be 4)
Cylinder	3
Cone	1

The model **sphere_cone_volume.scad** generates six pieces: a half-sphere (better known as *hemisphere*), a cone, and a cylinder that all have the same radius, **r** (in mm), defaulted to 25mm, and molds of each of those objects. The half-sphere, cone, and cylinder all fit into their respective molds. We didn't make a full sphere because it is too hard to 3D print and use as a mold.

You can adjust the value of **r** to get a set of bigger or smaller pieces. It is better to adjust the value of **r** in OpenSCAD versus scaling in your slicer since some detailed features of the molds, and clearances, might not scale well if you scale the whole model. If the fit of the pieces into their molds is too tight, increase the value of the **clearance** variable (defaulted to 0.3mm). You'll need it for the density assessment later in this chapter, so be sure to write down what value you used.

In Figure 8-7 we can see all six pieces. Notice that the cylinder has a little handle on one end. That's to help you get it out of the cylinder mold, which tends to get stuck otherwise.

Now you can put each model into its corresponding mold to see that they fit together and that the hollow molds have the same volume (within the clearances) as do the original shapes (Figure 8-8).

FIGURE 8-7: The cone, sphere, and cylinder molds (back row) and corresponding positive models (front row).

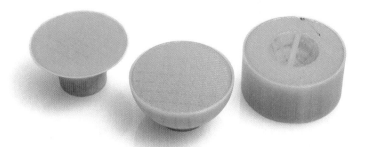

FIGURE 8-8: Positive models fitted into their respective molds

Let's try an experiment to see if it is true that the ratio of the volumes of a cone, half-sphere, and cylinder is 1 : 2 : 3 (when the radius of the cone and cylinder are equal to their height). First, fill the cone with something that won't pack down, like granulated sugar or salt. Fine sand will work, too. Powders like flour or powdered sugar pack down, so they will be less precise for this purpose.

Of course, water would be the obvious choice, but it's sometimes hard to get 3D prints to hold water around seams, depending on how well-tuned the printer is. The surface tension of water also makes it more difficult to fill to a precise level, and to avoid spilling. For these reasons, we suggest something granular instead. We show the process in Figures 8-9 through 8-14 using Himalayan pink salt. (Ignore the fact that we spilled a little - you will, too, and the differences will be smaller than can be measured.)

FIGURE 8-9: Fill the cone mold.

FIGURE 8-10: Pour the contents of the cone mold into the half-sphere, then fill the cone again.

FIGURE 8-11: Add the contents of the cone to what is already in the half-sphere. The half-sphere should just exactly fill.

FIGURE 8-12: Now, get your cylinder. Pour the contents of the half-sphere into it.

FIGURE 8-13: Then fill the cone a third time.

FIGURE 8-14: Pour the contents of the cone and the half-sphere into the cylinder (1 + 2 = 3). You will see that the cone + half-sphere fills the cylinder.

If the radius and the height of the models are not equal, the relationships get more complicated. We will see in the next section on pyramids that the relationship of 1:3 in volume between a cone and a cylinder holds as long as the cone and cylinder have the same base area and height as each other.

Once you finish this experiment, save the granular material you were measuring so we can find its density later in this chapter.

VOLUME OF PYRAMIDS

We would like to figure out the volume of a *pyramid*, which is a shape (like the Egyptian pyramids) that is similar to a cone, but with flat triangular sides (Figure 8-15). The base is a polygon.

Let's imagine that a pyramid and a cone have the same base area and the same height as each other, as is the case for the two in Figure 8-15. Can we relate their volume? Cavalieri figured out that, in fact, they are equal.

Imagine you have two planes that are parallel to each other. Let's say that

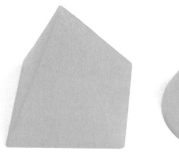

FIGURE 8-15: A pyramid (left) and cone (right)

one plane includes the base of the pyramid and cone, and the second plane is just where the cone and pyramid come to a point (that is, their cross-sectional areas go to zero).

We already said that the base areas of the two are the same, and they are the same height, so their cross-sectional area goes to zero at the same height as each other. We can also argue that since the cross-sections start the same and are continuously decreasing as you go up at the same rate, they will be equal everywhere. Thus, by Cavalieri's principle, they are equal. So we know now that the volume of a pyramid is ⅓ base area * height. Note that it doesn't matter how many sides the pyramid has.

FIGURE 8-16: Pyramid and cone with the same volume

There's another way to think about it, too. In Chapter 7, we saw that as the number of sides of a circumscribed or inscribed polygon goes up, it approaches being its circle. So we can think of a cone as a pyramid with a very large number of sides. As long as we are holding the cross-sectional area constant, the volume will stay constant as we increase the number of sides. In fact, 3D models like the ones we used above can't have perfectly smooth curves, because of how the file format works. What look like circles are actually regular polygons with straight sides, the sides are just so small that you can't see them (as we saw in Chapter 3).

Try creating a pyramid and cone with the same volume, as we saw in Figure 8-16. Fill one of them to just level with salt or sand (Figure 8-16). Now pour the material into the other one. The same, aren't they? These were created with **pyramid.scad** volume (**v**) of 25,000 cubic mm and a height (**h**) of 50 mm.

COMPARING VOLUMES

Now that you you can calculate the volume of cones, pyramids, and prisms, try printing out several that are the same volume to help your intuition. Here's a summary of what we have learned so far in this chapter.

Shape	Volume
Sphere	$\frac{4}{3}\pi r^3$
Cylinder	$\pi r^3 h$ (or base * height)
Cone	$\frac{1}{3}\pi r^2 h$ (or $\frac{1}{3}$base * height)
Prism	base * height
Pyramid	$\frac{1}{3}$base * height

Try printing a prism that is the same volume as a cone, but 1/3 the height. The base area should then be the same. A prism with the same base and height as a cone will have 3 times the volume.

IF YOU DON'T HAVE A 3D PRINTER

If a 3D printer isn't available, you can make these parts out of things you have lying around, although it will be challenging to get them precisely in the ratio here.

Cylinders are easy to find (straight-sided cups or glasses, or soup cans will work). Cones are easy to make out of paper and should self-straighten into a better cone when filled, since the correct shape is the maximum volume for its surface.

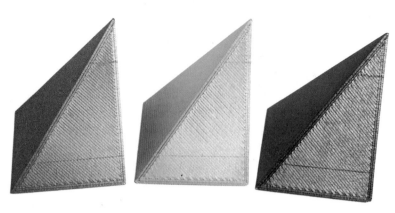

FIGURE 8-17: The three pyramids that together have the same volume as a cube

A hemisphere is a little tricker. Ping pong balls cut in half would be good, or the bottom of a plastic easter egg is probably close enough. Craft stores also have holiday ornament balls that split in two.

Of course, in any of these cases, accurately measuring the internal dimensions will be the hard part, and you won't get them to all be equal radius and height like the printed ones. Also, for most of the hemisphere options, you will need to measure the base radius and height separately, since they may not be equal (which means results will be a bit off). In this case, you'll have to use ⅔ of the base area times the height for the volume of your hemispheroid.

CAVALIERI'S PUZZLE

Earlier in this chapter, we said that the volume of a cone or a pyramid is 1/3 the base times the height. To try to convince you a bit more than we can with moving sand around between models, let's try out a puzzle. We can show an example of a cube that is created from three identical pyramids, with the model **3_pyramid_puzzle.scad**. Each of the pyramids has a base equal to a side of the cube, and so collectively they should add up to the cube (Figure 8-17).

The only parameter in 3_pyramid_puzzle.scad is the length of a side of the cube, the variable **size,** in mm. Since relative dimensions are crucial, if you want to scale this model, you can do so by changing **size**, or by scaling it (uniformly) in your 3D printer's slicing program. If you scale one axis more than another, the pieces won't fit.

Try rearranging these pyramids into a cube to prove that it works. As a hint, note that the square sides need to be on the outside, and some sides of the

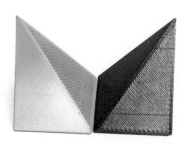

FIGURE 8-18: First, make sure you have a square side down on the first block, since that will be the base of the cube.

FIGURE 8-19: Next, turn the second piece so that it, too, has a square side down, and one of its big sloping triangles lined up with the big sloping triangle of the second side.

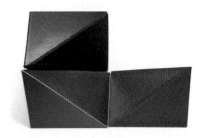

FIGURE 8-20: Now, rotate the second pyramid so the two sloping triangles touch. The second pyramid's base becomes another side.

cube will be half one pyramid and half the other.

If you can't quite get it to work, the solution is shown in Figures 8-18 through 8-21.

DENSITY

The density of a material is the mass of a standard volume (like a cubic centimeter) of that material. For example, one gram of water has a volume of one cubic centimeter, so the density of water is one gram per cubic centimeter, or 1 g/cm³. That's not accidental — the metric system was originally defined that way, although it was later adapted to more precise standards.

FIGURE 8-21: Take the third pyramid, and line it up so that its two sloping triangles will fit into the valley made by the first two pyramids. Then rotate the final pyramid into place to make the cube.

MASS v. WEIGHT

You'll note that we talked about mass and not weight. In everyday use, we use these somewhat interchangeably. Technically, though, the mass of an object (measured in kilograms or grams, in the metric system) is the same everywhere. The *weight*, which is the force exerted by gravity on an object with a particular mass, is measured in Newtons in the metric system, and in pounds in the Imperial system. The mass is a property only of the object, but the weight depends on both the object and the gravity in the environment where it is placed (such as on the Earth's surface or in space). Since you are (presumably) going to do these experiments on the Earth's surface, the distinction may seem a little fussy. However, density is in units of mass per cubic volume (not force per cubic volume), and we'll use the more precise term here.

FIG. 8-22: Empty cylinder on scale   **FIG. 8-23:** The full cylinder on the scale

MEASURING DENSITY

We just used salt, sugar, or sand to explore volume. Sand, in particular, will have a different density depending on what types of rocks were ground down to create it. Salt or sugar will vary a bit too, based on any impurities, whether they have absorbed water, and the like. You'll have to measure a known quantity first and then go from there. You could also try Play-Doh or similar substances if you can smooth off the top evenly enough. To do this, you'll need to put in more than enough to fill the space, then cut off the excess (dental floss works well for cutting Play-Doh).

Here is how you can measure density. First, take the hollow cylinder mold model. Density is often quoted in grams per cubic centimeter, so let's first get the radius in centimeters. One centimeter (a hundredth of a meter) is 10mm (a thousandth of a meter). If we used the default of r = 25mm, that's 2.5cm. The volume inside this mold is πr^3, or $\pi * (2.5cm)^3 = 49\ cm^3$.

Next, use a postal scale to measure the mass of the cylinder model with nothing in it (Figure 8-22). That is called the "tare weight" of the container. If possible, measure the mass in grams. (For some reason, no one says "tare mass", even though they should.) Ours, printed in PLA at the default radius of 2.5cm, had a tare weight of 8.00g. Most digital scales will have a tare function that allows you to place your empty container on the scale and zero it, so that you are only measuring the contents. We'll show the process without using this function, in case your scale doesn't have one.

Next, fill the cylinder with the sugar, salt, sand, or whatever else you used in the earlier experiment. We used Himalayan (pink) salt. Use the postal scale to measure the mass of the cylinder model plus the material. When we tried it (Fig. 8-23), we got 63.89 grams for the cylinder plus salt.

To get the mass of the salt alone, subtract the tare weight of the cylinder from the total.

In our experiment, it was 63.89 - 8.00 = 55.89g.

To get the density of the salt alone, divide the mass by the volume. In our experiment, 55.89g / 49 cm³ = 1.15g/cm³

This is what is called the **bulk density**, which allows for air that gets trapped between the grains of salt. (If you could somehow stuff salt grains into one giant salt crystal, it would be denser than the loose granular salt. For example, you could dissolve it in water and let the material dry out and re-crystallize into one giant crystal.

Try the experiment we just described. Your answers will be a little different depending on the density of your 3D print, how finely your salt was ground, and how carefully you can measure out your salt.

DISPLACEMENT

Imagine that you have a glass of water that is full right to the brim. Now suppose you slide a few pennies into the glass. Some of the water will flow out, but how much? The water in the glass had to make room for the pennies, so enough water to equal the volume of the pennies would splash out. It doesn't matter how much the pennies weigh, as long as they will sink in water (that is, if their density is greater than that of water, which is 1g/cm³). The pennies take the place of, or *displace,* the water, so we call this phenomenon *displacement*.

You can try this out for yourself. Get about 10 circular coins (like pennies, if you are in the United States). Find their diameter and the height of the stack of 10 with a good ruler, and use the formula for a cylinder to get the volume they take up. The US Mint says that a penny is 1.905cm in diameter (and thus 0.9525cm in radius) and 0.152cm thick, so 10 of them should be 1.52cm thick. The volume of a cylinder is pi times the radius squared times the height, or 3.14149 * (0.9525cm)³ * 1.52cm³ = 4.33cm³, which is also equal to 4.33ml.

FIGURE 8-24: Small measuring cup before adding pennies, with 30ml water

FIGURE 8-25: After adding pennies (water level about 35 ml)

Now, partially fill a measuring container (Figure 8-24). The tall and thin cylindrical measuring cylinders used in labs, called graduated cylinders, would be best. A kitchen measuring cup marked in mL or cm^3 would be good too. Add the pennies (Figure 8-25). In our case, it looks like it went up about 5ml, which is pretty good compared to our calculated value of 4.33ml given that our measuring cup is graduated in units of 5ml.

How much did the water level go up when you tried it? Given that 1ml = $1cm^3$, see if the volume displaced is the same as your computed volume for the stack of coins. Was it? (Within the error of your measurements, of course).

MEASURING VOLUME WITH DISPLACEMENT

One application of this process is to measure the volume of something that is a weird shape, like a crown or a statue. As we mentioned in the introduction to this chapter, Archimedes was asked by his king to establish that a crown a craftsman had just made for him was made of pure gold, as required. Gold is very dense (about 19.3g/cm^3). How might we use displacement to out our unethical crown maker?

Presumably, a shady artisan wouldn't want to get caught, so he would make the easy-to-measure characteristics of the crown the same as the real article. A balance that could measure reasonably well that two things had the same mass would have been available, so let's assume the mass of the crown was the weight of the gold that was paid for. However, other, less-expensive metals have lower densities (like silver, at about 10.5g/cm^3). That

The Questionable Crown-Maker

Let's calculate an example of how much different displacement of a crown of mixed metals would be versus one of pure gold.

Let's say our artisan decided he could get away with making the crown 80% gold (by mass) and 20% silver, as long as its mass was the same as the customer expected from the mass of pure gold they paid for.

Just as a rough guess, imagine the crown was about 10cm in radius, about 0.1cm thick, and about 5cm high. We could find that volume by subtracting a cylinder 9.9cm in radius from one 10cm in diameter, leaving a simple crown sort of like a giant napkin ring with straight sides. A crown would be fancier than this of course, but we can approximate to see what the numbers look like.

Volume of a cylinder = $\pi r^2 h$
Volume of a tube = $\pi h (r_{outer}^2 - r_{inner}^2)$

For our example, volume of cylindrical crown = $\pi * 5cm * ((10cm)^2 - (9.9cm)^2) = 31.3 \ cm^3$

A crown of this volume made of pure gold would be $31.3 \ cm^3 * 19.3 \ g/cm^3 = 604g$

However, a crown made of 20% silver would take up a different volume if it had the same mass, since a gram of silver takes up nearly twice the volume of a gram of gold.

Let's see how much of this crown is actually taken up with gold, if we are keeping the weight the same. 80% of the volume of the 31.3 cm3 pure gold crown would be 25.0 cm^3, and 80% of the 604g mass would be 483 g.

That means that the remaining 20% of the mass of the whole crown, or 121g, would be silver. The density of silver is about 10.5g/cm^3. We can calculate the volume of the silver by dividing the mass by its density.

Volume of the silver = $121g/10.5g/cm^3 = 11.5cm^3$
Total volume of the crown = volume of gold + volume of silver = $25.0cm^3 + 11.5cm^3 = 36.5cm^3$ or about 5.2cm^3 more than the pure gold crown.

means that an all-silver crown would have to have nearly twice the volume as an all-gold one to weigh the same. A smaller fraction of silver mixed with the gold, though, would have been trickier to detect. See the sidebar for the details of the calculation.

Archimedes could have tested this by putting a known gold block of the right mass underwater, checking the water level, and then looking to see if he got a different result when he lowered the crown into the water. Of course, if the crown was more delicate or a smaller fraction of the gold was a different metal, then the difference might have been undetectable by the technologies of the time. Air bubbles, either trapped in a hollow space within the metal as it was being shaped or captured on the surface of the crown as it was being submerged, would be added to the apparent volume.

FINDING DENSITY FROM DISPLACEMENT

If you want to find out the density of something oddly shaped, like a handful of screws, you can proceed as follows.

First, weigh the object(s) of interest on a postal or kitchen scale, and write down the mass (we'll assume we are using grams). In our case (Figure 8-26) our screws weighed 24.29g.

Fill a graduated cylinder, or at least the most precise measuring cup you have, and fill it with enough water to cover the object you are analyzing (but don't put the object in yet). A tall, narrow container can be read more precisely because adding a small amount of volume will make the water level rise more. Note how much water is in the measuring cup (30ml in our case, in Figure 8-27).

Now put in your objects and see what the difference in volume is (being careful with the units!) In other words, see how much the water level went up. If your container is marked in ml, that is the same as cm^3. If it is marked in cups, 1 US cup equals $236.588cm^3$ so you'll need to multiply your result by that to get cm^3. In our case (Figure 8-28) the water rose by about 4 ml.

The mass divided by the volume *difference* (with and without the object in the water) will give you the average density of the object. With precise enough measurements, you might be able to look up densities of common materials and deduce what your submerged object might be made of. This will work well for things like metal, as long as you are putting in something big enough so that it is easy to see. In our case with our screws of an

FIGURE 8-26: Screws on a postal scale

FIGURE 8-27: Measuring cup before adding screws

FIGURE 8-28: Measuring cup with screws

unknown steel alloy, we got $24.29g/4cm^3$ or about $6g/cm^3$. Steel is about $8g/cm^3$, so this is a little inaccurate because of our limited ability to measure water volume within a milliliter, and the tendency of air to get trapped on the screws. Adding more (identical) screws will give you larger mass and volume numbers, which may be easier to measure accurately.

This method will require care for anything porous that absorbs water. If you completely saturate the porous material, you'll be measuring its density when it is full of water, not as it would be on land. Also, this technique only works for things that are heavier than water. In the next section, we'll see how to measure something lighter than water.

ARCHIMEDES' PRINCIPLE

Despite the charm of the crown story, Archimedes is usually remembered for his realization that the situation is different for a body that is *floating* in (or on) a fluid. For our counterfeit crown test, we assumed the crown sank and was resting on the bottom. But floating is a little more complicated to think about.

Why does a boat float at all? Some ship-building materials, like oak and other types of wood commonly used in ships, are less dense than water, and so it's not all that surprising. However, what about metal ship hulls, including ones that are carrying many tons of cars or electronics across the Pacific? Archimedes realized that a floating body will displace an amount of water (or other fluid) of equal *weight*. Today, this is called *Archimedes' Principle*. Note that the word "weight" is appropriate here, since we are measuring a force. In zero gravity, the boat, water, deck chairs, and passengers would all be floating around in a big mess. Gravity is required for orderly floating.

BUOYANCY FORCE

Assuming we are on a planet, anything placed in water is being pushed and pulled by several forces. First, gravity pulls down on it. If its average density is higher than that of water ($1g/cm^3$), then it sinks. (It's not an accident that $1cm^3$ of water has a mass of $1g$; the metric system was designed that way since it is so often more convenient compared to needing to divide by 62.4 lbs/ft^3 all the time.) When something sinks, it has to push water out of the way. You can think of this as the water exerting a force equal to the weight of the water pushed out of the way, called *buoyancy*. If the density of an object is exactly the same as that of water, it will displace exactly the same amount of water as itself and will be fully submerged, but will not rise or sink unless some other force acts on it.

If, however, its *average* density of an object is lower than that of water (say, a metal-hulled ship with thousands of cars on board but a lot of air in the hull) it will float. But some of it will be below the surface of the water. Ships are rated by their maximum displacement, which is how much the boat, fuel, cargo, passengers, and anything else can weigh in total before the ship sinks. Or, in other words, the weight of the water pushed aside by the underwater part of the ship.

Let's take a step back now to realize that we can deduce the volume of an object if it sinks on its own, but not its weight. We can find the weight of something if it floats, but not its volume. (Assuming, that is, that we are deducing these quantities purely from the amount of water displaced.)

	What I have to measure another way	What I can deduce
Floating ship	Volume	Mass
Sunken treasure	Weight	Volume

However, if you have something that does not float, you can drop it in water, find out its volume, then put it in a very lightweight container that will float to get the weight. The average density of the container, the air, and the object will let it float, you will be measuring the combined weight of the enclosed air and the container as well as your object of interest.

If we are in a known gravity field (like the surface of the earth), the mass and weight measurements can easily be deduced from each other, as we noted in the Density section of this chapter. Once we have both mass (or weight) and volume, we can calculate the density to find out if our crown is made of gold or something else. We'd find the measurement by looking up the density of candidate materials or calculating likely mixtures until we got a match.

When we talk about objects floating in "a fluid" we can also mean floating in the air, like a blimp or a hot-air balloon. Hot air is less dense than cold air, so heating the air in a hot-air balloon will let the

FIGURE 8-29: Measuring cup with just water

FIGURE 8-30: Measuring cup with "boat" and ring

FIGURE 8-31: The ring's displacement

buoyant force become bigger than gravity, which makes the balloon rise.

It's obvious that hot-air balloons need to be light to fly. However, now that we understand ship displacement, we see how important it is for ships to be light, too. Otherwise, the average density for the ship plus a useful amount of cargo would get higher than the density of water (and would sink). Wood is significantly less dense than water (oak is around 70% the density of water, plus or minus 10%) and therefore has been a successful shipbuilding material for millennia.

For ships, it also matters whether you are in freshwater or seawater. Seawater is denser than freshwater, and so a ship will displace less seawater than it would freshwater. Some large cargo ships have a horizontal line, called a Plimsoll line, painted on their hull to mark the waterline under various conditions and guard against overloading.

BUOYANCY EXPERIMENT

It is tricky to measure displacement unless the object is big. 10mL is usually just a fine line on a measuring cup, and seeing differences any smaller than 10mL is hard without equipment beyond what most people will have at home. With that said, let's try to see if we can measure the density of a gold ring with a precise measuring cup.

First, we will fill the measuring cup with enough water to float a ring in a soda bottle cap, which we will use as a miniature boat (Figure 8-29). Now, we will carefully take a gold ring and put it in a bottle cap. It just barely floats. The water rises about 10mL above where it was before we put in the ring (Figure 8-30). That would imply that the collective mass of the bottle cap, the ring, and the air around them should be about 10g, since each displaced mL is equal to 1g of water.

We can then check our number by putting the bottle cap plus ring on a kitchen or postal scale and seeing if we get roughly the same number. The bottle cap plus the ring was 9.79g. The ring alone was 7.79g.

When we let the ring sink (Figure 8-31) it displaced less than 1 mL, so that would imply that the density of this ring (which is marked as being composed of 14K gold) would be something more than $8g/cm^3$. The density of 14K gold (which is defined as being 58.3% gold, or 14/24 gold; 24K gold is essentially pure gold) is about $13g/cm^3$. We can see it is pretty challenging to do these measurements for something small with home equipment; if you happened to have a big block of gold lying around, it would be a lot easier to try this measurement.

SUMMARY AND LEARNING MORE

Volume, density, and displacement have enormous practical value, and this chapter got you started with all three. Now that you have some formulas for finding the density of basic shapes, you might try estimating values of more complicated ones by imagining them as two or more basic ones smushed together, or consider how you might use Cavalieri's Principle to find a simpler equivalent to a complex volume. For example, the volume of ice cream in a cone plus a scoop might be a half-sphere plus a cone. If your local ice-cream place is particularly generous, It might be closer to a full sphere plus a cone, or it may be a cylinder, depending on the type of ice cream scoop they use. Wikipedia has good articles under "Volume," "Density" and "Displacement (ship)" that parallel the ideas in this chapter.

Shipbuilding is another area to explore, and we encourage you to play around with floating plastic cups, weighted with pennies, in a bowl of water to get a sense of how displacement (and stability of a floating object) work.

You might also explore and experiment with floating objects more generally, like a 3D printed part with infill and see if you can think about what their average density might be, based on how much of the object is below the waterline. Calculating this is complicated for a 3D print, since the print has outer shells and infill. The density of PLA is around $1.24g/cm^3$, although it can vary with additives.

You can also try to estimate the percentage of an iceberg that is underwater. For example: the density of freshwater ice is $0.92g/cm^3$, and seawater is about 1.0273g/cm³. The ratio between the two is 0.92/1.03 or 89%. Therefore about 89% of an iceberg would need to be underwater (in seawater) for the displacement to balance. Our book, *3D Printed Science Projects, Volume 2* from Apress (2017) has a chapter on snow and ice which describes how to create a 3D printed iceberg model.

CHAPTER 9
SURFACE AREA AND NETS

Surface area is the sum of the areas of all the faces (or curved surfaces) that make up an object. You can think of it as the amount of wallpaper you would need to completely cover the surface. How might you go about covering a curved surface smoothly with wallpaper? It gets a little tricky. Being able to calculate the surface area of an object comes in handy if you need to do something to the outside of a shape, like painting or covering with fabric. In the last chapter we figured out how much material might be in a container; in this one, we figure out how much material it would take to cover the outside of it.

One way to visualize surface area is to imagine flattening out the surface of a 3D object, as if it were made of paper and you made strategic cuts to lay it out in 2D. The resulting 2D surface is called a *net*. Any 3D shape made up of flat faces can, in principle, be flattened into a net. Curved surfaces like you would find on a cylinder or a cone can also be unrolled to produce a flat net, but surfaces that curve in more than one direction, like a sphere, can only be approximated on a flat surface. This is why flat maps of the Earth's surface will always have some distortion. For some cases that can be flattened, though. This chapter will show you how to create nets, with 3D prints or paper.

You can approach reading this chapter in one of two ways. If you just want to get a general sense of how various 3D solids can be created from 2D faces, you might want to skip the calculations and just create and explore models. If you want to be able to get the value of surface area for different shapes, though, we have included the details, which will require you to brush up on the Pythagorean Theorem and some trigonometry (Chapter 6). If you want to check you did the calculation right, the models also display the numerical value of surface area for the given set of parameters during the model rendering process. But before we get too far ahead of ourselves, let's understand a bit more about nets.

3D Printable Models Used in this Chapter

See Chapter 2 for directions on where and how to download these models.

platonic_net.scad
Produces nets for the Platonic solids

pyramid_prism_net.scad
Produces nets for pyramids and prisms

cone_cylinder_net.scad
Produces nets for cones and cylinders

Other supplies for this chapter
- A ruler
- Optionally, some thin fabric (like tulle)
- Optionally, supplies for gluing onto fabric (see section "Gluing Instead")

NETS

The German artist Albrecht Dürer was famous for engravings and woodcuts. He wrote a book called *Four Books of Measurement,* published in 1525, in which he illustrated Platonic solids and other shapes as 2D cutouts, and appears to be the first person to publish the use of nets and other speculations on

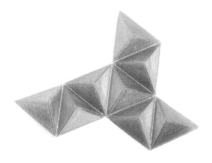

FIGURE 9-1: Octahedron net (open)   **FIGURE 9-2:** Octahedron net (folded)

perspective. (One of his engravings, *Melencolia,* currently in the collection of the Metropolitan Museum of Art in New York, shows a thinker looking unhappily at an irregular polyhedron. Any mathematics student will appreciate the sentiment.)

We have rethought the traditional, 2D, paper version of a net to make a more tactile version that is a little easier to assemble and handle. Our take on nets turns each flat face into a pyramid. The tops of these pyramids come together as the net folds up. The pyramids aid in both tactile counting of faces and also in folding the faces to the correct angles. The flat sides (on the bottom of the model in the open version) are what you will count up to get the surface area. For example, here we see the net of an octahedron open (Figure 9-1) and folded up (Figure 9-2).

Nets (other than approximations) can't be created for spheres since there is no way to cover these shapes with flat paper. Cones and cylinders do each have a relatively simple paper net (but not a 3D printable one). We'll describe the parameters for each one in the relevant section of this chapter.

3D PRINTING THE NETS

The OpenSCAD models for this chapter can create a 3D print that will fold on hinges. Unlike many of the other models in this book, they are a little tricky to 3D print. They have a thin *living hinge* (that is, a hinge printed of the same material as the rest of the print) that we print as a bridge over the gap between the polygons.

The variable **hinge** (in mm, defaulted to 0.201) is used to control how thick these living hinges are printed. These should only be about one or two layers thick. Making them thicker will make them stronger, but also more difficult to bend, and more likely to snap instead.

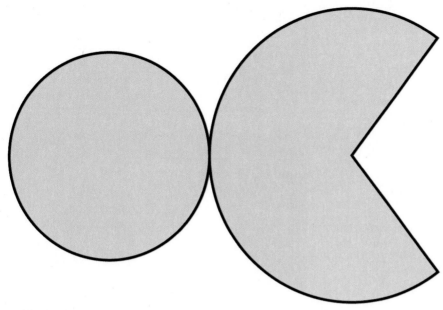

FIGURE 9-3: Cone net

We have had good luck using PLA, 0.1 and 0.2mm layer heights, and printing with 0% infill (that is, hollow). Some formulations of PLA are more brittle than others, and we've seen some (like HTPLA) snap apart on the first bend. You may have to experiment a little, and the users will need to be gentle with the prints. You might have an easier time with PETG if your 3D printer can use it.

If the prints do break at the hinges despite your best efforts, you can always use some tape to replace the hinge. (You'll probably want to put tape on both sides.) You might find that you need a bit of strategic tape to hold the folded print together, too. You may have to experiment a little with tape you have on hand until you find one that will stick to the layer lines.

PRINTING THE NETS ON PAPER

If you want to use more traditional paper nets, OpenSCAD can also export a 2D version. Change the variable **linewidth** in any of the nets models to the line thickness you want, in mm. Then render the model as usual. You can use the **File > Export** menu to select saving as an .svg file. You can then open a new browser window, drag the .svg file into it, and send the resulting image to a paper printer.

You can also export to a .dxf file for a laser cutter, but you don't really want to cut anything except the outer boundary of the net. To make that work, you will

```
Loaded design '/Users/joanhorvath/Dropbox/Geometry/2020-UpdatedModels/CommentedNew/cone_cylinder_net.scad'.
Parsing design (AST generation)...
Compiling design (CSG Tree generation)...
ECHO: "surface area = 3033.79mm^2"
Rendering Polygon Mesh using CGAL...
Geometries in cache: 10
Geometry cache size in bytes: 33136
CGAL Polyhedrons in cache: 0
CGAL cache size in bytes: 0
Total rendering time: 0 hours, 0 minutes, 0 seconds
Top level object is a 2D object:
Contours: 3
Rendering finished.
```

FIGURE 9-4: OpenSCAD console screenshot, highlighting the surface area

FIGURE 9-5: Tetrahedron net, open

need to edit the file in Adobe Illustrator or a similar program before trying to laser cut the net. Be careful if you decide to scale them on printing out that they don't cross page boundaries or scale differently from side to side and top to bottom of the page. In the case of the cone and cylinder nets, paper is the only option since flat 3D prints won't work for curved surfaces.

For most polyhedrons, there are many possible ways to break up the surface into a net, since there are multiple ways to connect up the sides to make them convenient to assemble. You may see alternatives in other publications and you might want to experiment with those, too.

FIGURE 9-6: Tetrahedron net, folded

3D PRINTING THE NETS ON FABRIC

One final alternative is to 3D print the nets on fabric. We describe this in detail at the end of this chapter. This solves the problem of the hinges being brittle or cracking, since we have fabric as the hinge instead. This is an advanced 3D printing technique, though, so read through that section carefully before embarking on the project. We'll also discuss gluing the nets onto fabric in that section, as a more crafty alternative.

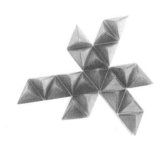

FIGURE 9-7: Icosahedron net, open

CHECKING ANSWERS

When you render each model in this chapter in OpenSCAD, it computes the surface area for the given parameters. Look in the "console" window at the conclusion of the render to see it (Figure 9-4). You can check your answers in each of the assessments that way. Be careful to check what variable each model is using (apothem, radius, or side). Values are in square mm. If you can't find the Console window, go to OpenSCAD's **View** menu, and be sure that the console window is checked.

FIGURE 9-8: Icosahedron net, folded

PLATONIC SOLIDS

Let's figure out the surface area of the Platonic solids we created in Chapter 3. To do that, we can look back at the area of inscribed and circumscribed polygons in Chapter 7 where we compute the area of a regular polyhedron. Consider printing several of the solids from **platonic_net.scad** to follow along with the geometrical reasoning here.

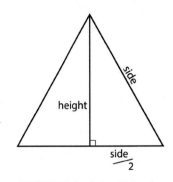

FIGURE 9-9: Equilateral triangle

MODIFYING THE MODEL

The model **platonic_net.scad** creates nets for Platonic solids. It has these parameters (and default values):

- **faces = 4;**
    - Number of faces
- **edge = 10;**
    - Length of an edge of face, mm
- **linewidth = 0;**
    - Cut/fold line width for 2D export. Set to zero for 3D.

The Platonic solids have faces that are either equilateral triangles, squares, or regular pentagons. That means we need to find the area of each of those fundamental shapes, and then add up those areas to get the surface area of the solid. As we learned in Chapter 3, the apothem is the distance from the center of a regular polygon that makes a right-angle intersection with one of its sides.

EQUILATERAL TRIANGLE FACES

Three Platonic solids have equilateral triangle faces: the tetrahedron (4 faces, Figures 9-5 and 9-6), the octahedron (8 faces, back in Figures 9-1 and 9-2), and the icosahedron (20 faces, Figures 9-7 and 9-8). The model takes as input the apothem, in mm, since that simplifies the math to make the net. The apothem of an equilateral triangle is just half its height.

To calculate the area of the triangles that make up the faces of the Platonic solids that have equilateral triangle faces, we can use the Pythagorean Theorem to find the height (h) of the triangle as a function of the length of an edge, which we'll call s, as shown in Figure 9-9.

$$(s/2)^2 + (h)^2 = s^2$$

Which, if we gather up the terms in s, becomes

FIGURE 9-10: Cube net, open

FIGURE 9-11: Cube net, folded

$h^2 = 3s^2/2$, or
$h = \sqrt{3}\, s /2$

The area of each equilateral triangle is 1/2 base times height (twice the apothem), so we get:

Surface area of each face = $(1/2) * s\ \sqrt{3}s /2 = \sqrt{3}s^2 /4$

You would then multiply this by the number of faces to get the surface area of the tetrahedron, octahedron or icosahedron. For example, the icosahedron has 12 sides, so the surface area of the icosahedron is:

Surface area = $12 * 1.7205s^2 = 20.646s^2$

SQUARE FACES

The only Platonic solid with square faces is a *cube*. A cube has 6 faces, and so we would set **faces = 6.** Figure 9-10 shows the net open, and Figure 9-11 folded into a cube. The surface area of a cube is just 6 times the area of a face, which is in turn just the edge length squared.

Surface area of a cube = $6s^2$

PENTAGONAL FACES

A dodecahedron (12 faces) is the only Platonic solid with pentagonal faces. In Chapter 7 we saw how to compute the area of an n-gon by breaking it up into triangles. As we worked out in Chapter 7's section "Area of Inscribed Polygons", we can cut the pentagon into 10 small right triangles, each of which will have a central angle of 36°. The height of this triangle is the apothem of the polygon (Figure 9-12).

$\tan(36°) = (s/2) / h$, or

$h = s/(2 * \tan(36°))$

Area of the pentagon $= (1/2)$ base * height * 10
$= (1/2) s/2 * (s/2)/\tan(36°) * 10$
$= 1.7205 s^2$

And so, the surface area of the dodecahedron is just the area of the pentagon * 12. Figure 9-13 shows a dodecahedron net open (12 faces, pentagonal sides) and Figure 9-14 shows it folded.

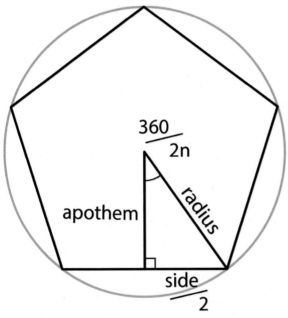

FIGURE 9-12: Anatomy of an inscribed pentagon

FIGURE 9-13: Dodecahedron net open

FIGURE 9-14: Dodecahedron net folded

CHECKING SURFACE AREA OF PLATONIC SOLIDS

In summary, then, if we want to compute the surface area of a Platonic solid that we have printed out as a net, we would set the parameter **faces** to be the number of faces, and the parameter **edge** equal to the length of an edge (side of the face).

	Shape of faces	Number faces (**faces**, in the model)	Area of face, as a function of length of an edge (**edge** in the model, s here)	Surface area of polyhedron, as function of s (**edge**)
tetrahedron	equilateral triangle	4	$\sqrt{3}\ s^2/4$	$\sqrt{3}\ s^2$
cube	square	6	s^2	$6s^2$
octahedron	equilateral triangle	8	$\sqrt{3}\ s^2/4$	$2\sqrt{3}s^2$
dodecahedron	regular pentagon	12	$1.7205\ s^2$	$20.646\ s^2$
icosahedron	equilateral triangles	20	$\sqrt{3}\ s^2/4$	$5\sqrt{3}\ s^2$

Print (3D or 2D) several of the Platonic solid nets from **platonic_net.scad**. Count up the faces and note the shapes of each face, and see that they match the entries in this table. Note that these relationships only work if all the faces are regular polygons identical to each other. Otherwise, the adjacent faces would have to distort if one face got bigger in any dimension.

RIGHT PRISMS

A right prism (that is, one whose top and bottom are parallel to and directly above each other) is a general case of the cube, which we covered under the Platonic solids. As with a cube, to get the surface area, determine the top and bottom surface area, and then add that to the rectangles making up the sides. We will need to calculate the area of the top and bottom by breaking it up into triangles (as we did for the Platonic solids).

Each rectangle making up a side will have an area of length times height. There will be n of them, if it is an n-sided prism. The total surface area will be:

Surface area = 2 * (area of the top) + n * (area of the sides).

MODIFYING THE MODEL

The model **pyramid_prism_net.scad** prints prisms and pyramids. It has the following parameters, which are set to defaults as follows:

- `r = 20;`
  - The radius of the base of the prism, in mm
- `h = 30;`
  - The height of the prism (or pyramid), in mm
- `sides = 5;`
  - Number of sides, not including the base
- `pyramid = false;`
  - Set this to false for a prism, true for a pyramid
- `star = true;`
  - If true, it arranges the parts in a star-like arrangement, with the base in the center.
- `linewidth = 0;`
  - cut/fold line width for 2D export. Set to zero for 3D
- `baselayer = 1;`
  - height of print below (outside) the hinges (in mm)
- `printheight = 500;`
  - total height in mm - set to less than h to have a truncated pyramid or cone on the inside of the print
- `hinge = 0.201;`
  - thickness of the hinges, in mm. Using a negative value will leave a gap

Here is an example of the default values, with a pentagonal prism. Figure

FIGURE 9-15: Pentagonal prism net, open

FIGURE 9-16: Pentagonal prism net, folded into the prism

9-15 shows the net laid flat, and Figure 9-16 shows it folded up into the prism.

To try this out, 3D print a prism with a square base using the **pyramid_prism_net.scad** model. Based on the values of **r** and **h** you input, estimate the surface area. Then try measuring the produced model and seeing how close you got. Remember that the radius of the base (**r**) parameter isn't a *side* of the base. The radius is the distance from a vertex of the base to the center. If we have a square base, the radius is the hypotenuse of a right triangle s/2 on each side. The Pythagorean theorem tells us that

$r^2 = (s/2)^2 + (s/2)^2$ which we can simplify to
$r^2 = 2s^2/4$ or further simplify to
$r = s/\sqrt{2}$

Thus, to get the side of the square to calculate area, you'll need to multiply the side by $\sqrt{2}$. If you want to explore other than square prisms, you can break the base polygon into triangles the same way we do with Platonic solids (although there we were working with the apothem, not the radius). Check yourself against what the model calculates when you run it.

PYRAMID

The surfaces of a pyramid are made up of triangles and the base is a polygon. The height of each triangle is equal to the slant height, and the base is the length of each side of the polygon making up the base. Let's assume that the base is a regular polygon (all sides the same) and that the vertex at the top of the pyramid is right over the center of the base, so we aren't dealing with any skewed sides.

As we saw when we calculated the inscribed polygon in Chapter 7, if there are n sides to our base polygon that are each of length s, we can break our base up into 2n triangles and calculate the base area that way. As the number of sides gets very big, the base starts to approximate a circle, and the pyramid approaches a cone. The model pyramid_prism_net.scad also prints nets for pyramids, with the parameter "pyramid" set to true.

Each of the 2n triangles has a base of s/2 and a height equal to the apothem of the polygon:

Apothem = $s/(2 * \tan(180/n))$

The area of the n-sided base with sides of length **s** is:

Base area = $n/(4 * \tan(180/n))$

Then we have to add in the area of n triangles on the sides of the pyramid. Each one has a base of length s, and a height equal to the slant height, which is the hypotenuse of a right triangle made by the height of the pyramid and the apothem.

Slant height $= \sqrt{apothem^2 + h^2}$

The area of each of these triangles is

Area = slant height $* s/2$

So the area of a pyramid with n sides is:

Surface area = $n * s *$ slant height$/2$ + base area.

Let's see what this is if we have a pyramid with a square base. In that case, the number of sides is 4, and $\tan(180/4) = \tan(45) = 1$. Figure 9-17 shows the net for this case laid flat, and Figure 9-18 shows it folded up into the pyramid.

Base area = $4s^2/4 = s^2$
Apothem = $s/2$
Slant height = $\sqrt{s^2/4 + h^2}$
Surface area = $s^2 + 2s\sqrt{s^2/4 + h^2}$

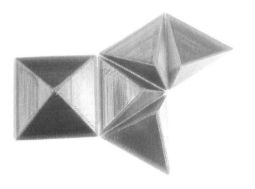

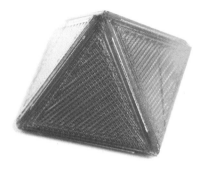

FIGURE 9-17: Square base pyramid net open

FIGURE 9-18: Square base pyramid net folded into a pyramid

Print out a pyramid with a square base using the **pyramid_prism_net.scad** model. Based on the values of **r** and **h** you input, estimate the surface area. Then try measuring the produced model and seeing how close you got. Remember that the radius of the base (r) parameter isn't a *side* of the base. The radius is the distance from a vertex of the base to the center. (For an accurate measurement, measure all the way across the base and divide by 2.) If we have a square base, the radius is the hypotenuse of a right triangle s/2 on each side. The Pythagorean theorem tells us that

$$r^2 = (s/2)^2 + (s/2)^2$$
$$r^2 = 2s^2/4$$
$$r = s/\sqrt{2}$$

So to get the side of the square to calculate area, you'll need to multiply the radius by $\sqrt{2}$. If you want to explore other than square prisms, you can break the base polygon into triangles the same way we do with Platonic solids (although there we were working with the apothem, not the radius). Check yourself against what the model calculates when you run it.

CONE

Technically a cone can be thought of as a pyramid with infinitely many sides. That's not easy to calculate until you move on to calculus, which has tools for such things. Instead, let's think about what it would look like to unroll the surface of a cone. In this case, because part of the net needs to curve around, we have to use a paper model of a cone's net.

Use the model **cone_cylinder.scad** to create a paper model. **File > Export > .svg** creates a file that can be printed on a regular paper print). Drag the .svg file into an empty page on a web browser, and print it from there. The model has these parameters and defaults:

- **r = 20;**
  - Radius of the base, mm
- **h = 20;**
  - Height of the base, mm
- **cone = true;**
  - Set to true to get a cone, false to get a cylinder.

It consists of two parts: the base (which is just a circle) and the curved part. We know what to do with the base; its area is just πs^2. The upper part, though, is a segment of a circle. Print out the cone net, and cut the shape out of the paper. Be careful to leave the two halves connected, though. Since we did not allow any extra space for overlapping, you might want to leave a few strategic tabs to use as anchors for tape or glue (Figure 9-19). To assemble, first bend around the partial circle to make the top of the cone (Figure 9-20). Then bend over the base and tape (Figure 9-21). Finally you will have the completed cone (Figure 9-22).

The slant height of the cone is the hypotenuse of a triangle one side of which is the height of the cone, and the other is the radius.

$$\text{Slant height} = \sqrt{r^2+h^2}$$

We can tell what wedge of a circle the upper part of the cone is equivalent to by thinking about the fact that if we had a full circle, the radius would be the slant height. However, the radius of the base of the cone is $2\pi r$. Thus the fraction of the circle we have is

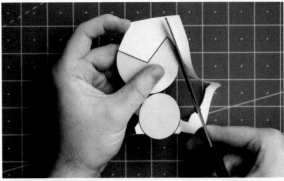

FIGURE 9-19: Cutting out the cone, leaving some tabs

FIGURE 9-20: Make curved part of the cone

FIGURE 9-21: Tape the base in place

FIGURE 9-22: Completed cone

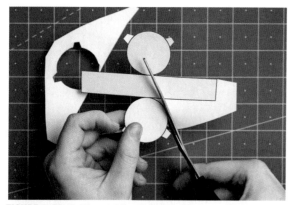

FIGURE 9-23: Cutting out the cylinder net

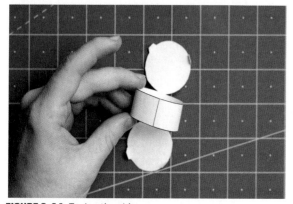

FIGURE 9-24: Taping the sides

FIGURE 9-25: Finishing off the cone.

Fraction of a full circle = $r/\sqrt{r^2+h^2}$
Area of the full circle would be $\pi\,(\sqrt{r^2+h^2})^2$

Area of the segment = area of the full circle * fraction of the full circle
$$= \pi\,(\sqrt{r^2+h^2})^2\,r/\sqrt{r^2+h^2} = \pi\,r\sqrt{r^2+h^2}$$

Surface area of a cone = area of the segment + area of the base $= \pi\,r\sqrt{r^2+h^2} + \pi\,r^2$

Try to think this through by printing (on paper) a cone net. Figure out, using the equations above, what its surface area should be. If you doubled the radius, what would that do to the surface area? Check yourself against what the model calculates when you run it, or check the "Answers" section at the end of the chapter.

CYLINDER

The surface area of a cylinder is easy to figure out if you think about the area of the label on a soup can. If you peeled the label off and flattened it out, it would be a rectangle $2\pi r$ by h, where r is the radius of the can and h is the height. The area of the base and top would be $2\pi r^2$ each. Thus the surface area of the cylinder is:

Surface area $\quad$ = side + 2 * base
$\qquad\qquad$ = $2\pi rh + 2 * 2\pi r^2$
$\qquad\qquad$ = $2\pi r(h + 2r)$

You can print (on paper) a cone net using the model **cone_cylinder.scad** as we just described for the cone. Set **cone = false** to get a cylinder. Cut the border of the shape out of the sheet of paper (being careful not to detach the two circles from the long strip. As with the cone, leave some tabs to help with assembly (Figure 9-23). Tape together the long strip to make the sides (Figure 9-24). Finally, tape on the top and bottom lids of the cylinder (Figure 9-25).

Print (on paper) a cylinder net. Figure out, using the equations above, what its surface area should be. If you doubled the height, what would that do to the surface area? Check yourself against what the model calculates when you run it, and look at the Answers section at the end of the chapter.

SPHERE

That the surface area of a sphere is $4\pi r^2$ is challenging to prove without calculus. However, some complicated proofs survive from Archimedes and others, which involve projecting the sphere onto a cylinder of equal radius. There is not a simple net for a sphere that conserves its surface area exactly, although you can imagine approximations if you think about ways to peel an orange and cut the peel into smaller and smaller pieces until it is as flat as you want.

FIGURE 9-26: Setup for hot-gluing onto fabric.

GLUING OR PRINTING A NET ONTO FABRIC

GLUING NETS ONTO FABRIC

The nets **platonic_net.scad** or **pyramid_prism. scad** are designed so that you can print out the separate pieces (without the hinge) and glue them onto fabric instead. You will need to be careful in the placement, though, so that the piece folds up correctly. For this method, set **hinge = 0** and **baselayer = 0**, so that you won't be pulling on the fabric when you try to fold the pieces together. (If you want a slight gap instead to allow room for heavier fabric, set **hinge = -1** for a 1mm gap, **hinge = -2** for a 2mm gap, and so on.)

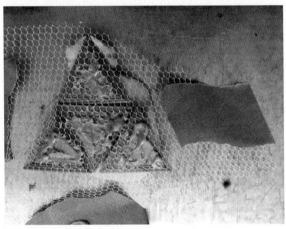

FIGURE 9-27: Hot gluing

We will describe how to use a hot glue gun to do this. You might experiment with other methods. First, we took a baking sheet and some clay, and arranged the model points-down in the clay so they wouldn't move around. This gives us a flat surface to glue on. Next, we taped down the fabric to the baking sheet (Figure 9-26).

Finally, we applied hot glue through the fabric. (We used colored hot glue to make it easier to see here, but regular hot glue will be less obtrusive.)

FIGURE 9-28: Finished with gluing

FIGURE 9-29: Finished net

Obviously, this only works if the fabric is loose weave or netting (Figure 9-27). As always, be careful with hot glue and don't burn your fingers! Take off the tape, and peel the net off the clay (Figure 9-28). Then cut off the excess fabric, and it will fold as shown (Figure 9-29).

FABRIC PRINTING PROCESS

It is possible, although tricky, to stop a 3D print part way up and insert fabric between desired layers of a print. People have been doing this to make costumes and other wearables; see, for instance, the work of David Shorey (www.shoreydesigns.com) or Billie Ruben (www.billieruben.info). You can visit either of their sites for ideas and techniques. In particular check out Billie's guide to 3D printing on fabric on her site, which has many detailed photos of the process.

Note that this is an advanced procedure, and you shouldn't attempt it unless you are very comfortable with your 3D printer hardware and software. The details are very printer-dependent, and we would recommend searching around a bit to find people who have done this with hardware like yours. We will, however, go over the general process and issues that arise so you will know how to set up the models.

You can use models from **platonic_net.scad** or **pyramid_prism.scad** for this process. First, set the parameter **hinge** to a negative number. Start with **hinge = -1** which will leave a gap where the hinge would be, and the fabric will be taking the place of the living hinge. A value of -1 will give a one-millimeter gap, -2 a two-millimeter gap, etc. One millimeter should be a large enough gap for thin, flexible fabrics like tulle, but thicker, stiffer ones may require a larger gap.

There are several ways to 3D print on fabric with a filament printer. One way is to attach fabric to the bed of your printer, either with small clips or tape, before you start printing. This has the advantage of being simple. However, it might confuse any automatic calibration your printer does at the beginning of a run. The plastic also might not stick to the fabric all that well, or the print may shift and fail if the fabric wrinkles or stretches.

What most people do instead is to print a few layers, pause the printer, clip or tape down the fabric, and then resume the print. If the fabric is netting or otherwise porous, the idea is that the plastic permeates through the fabric from above and bonds to the plastic below. It then solidifies around the threads of the fabric for better adhesion, so that the print doesn't just peel off.

It is often recommended to print 3 to 5 layers before pausing to add the fabric, but it's possible to do more. Pausing after you have finished the solid layers at the bottom of a print may result in a slightly weaker bond, but will avoid a lot of potential issues caused by trying to force too much plastic into the space already taken up by fabric. However you decide to print your base, you should ensure that the **baselayer** value is set to the height at which your pause will occur, so that your fabric hinge will end up in the right place.

PAUSING A 3D PRINT

Most popular slicers like Ultimaker Cura and PrusaSlicer can insert a pause between layers, either after a certain number of layers have been printed or once the printer has reached a certain height. This feature may be made available via a plug-in, extension, or post-processing script. They usually include options like moving the print head out of the way, displaying a custom message on the printer's control panel, and extruding a bit of material to prime the nozzle before resuming.

Alternatively, if your printer has a control screen and a "pause" (or "change filament") feature, you can use that to pause the print after you have seen the appropriate number of layers go by. If you have to resort to "change filament" you might have to fake out the printer by removing and replacing the same material, or perhaps take advantage of that to have the outer parts of the nets one color and the inner parts another.

If you run your printer via Octoprint or a printer-specific wireless interface, your manufacturer might have created a pause and restart functionality that way. Check your documentation to see how that might work.

For this method to work, you'll need some way of interacting with your printer while it's running, at least to tell it to resume after you've inserted the fabric. The appropriate pause codes and resume procedures will depend both on what printer you're using and what software you're using to control it. If there aren't obvious options for this in your software, you may want to go with the technique of putting down the fabric before the first layer is printed instead.

FABRIC TO USE

Use a thin fabric, and one that won't melt at the touch of the nozzle. Tulle works well, or other netting that will allow the plastic to attach itself through the fabric. Anything too thick will distort the print or make the nozzle get stuck, because there is no allowance for the fabric in the models. Always closely monitor a print on fabric to make sure the fabric isn't getting over-heated or caught in the nozzle.

ATTACHING FABRIC TO THE PLATFORM

You will need to somehow pin down your fabric to the print bed, either at the beginning or part way through the print. Depending on the geometry of your printer, you may want to use binder clips to attach it to the edges of the print bed, or some people have had good luck with using blue tape to attach it to the platform (using a cold print bed). During this process:

- Keep your hands away from the hot nozzle.
- If using clips, be sure that they won't be hit by the nozzle when it is moving (you'll need to bear that in mind when positioning the print on the bed during slicing).
- The fabric should be taut, with no wrinkles or inconsistent stretching. If the fabric is a little stretchy, you'll want to pull it pretty tightly over the platform.
- If you are taping the fabric down, the piece should be smaller than the platform. If you are clipping it, it will need to be a little bigger.
- Be sure that any excess fabric that hangs over the sides doesn't get close to any moving parts of the printer that it might get tangled in or wrapped around.

FIGURE 9-30: Octahedron on fabric

KEEPING A WARY EYE

Once you've stretched the fabric and started (or re-started) your print, hover over it for at least the first few layers and expect to have to start over a few times. It's easy for the fabric to pull up, get wrapped around the nozzle, or for the next layer not to print. It's always a good idea to stay near a print in progress, but we wouldn't leave a print on fabric alone until it completes. If things go badly, be careful unraveling the detritus, sigh, and try again (or perhaps hot glue will look better at that point). However, if it works, the results can be impressive (Figure 9-30).

SUMMARY AND LEARNING MORE

In this chapter we learned about surface area, and how to use nets to visualize them. We explored models to create nets for Platonic solids, pyramids and prisms, and cones and cylinders. These can be created on paper, with 3D prints, or in a mix of 3D prints and fabric. There are good discussions about surface area on Wikipedia, and several videos through the Khan Academy.

If you want to learn more about Albrecht Dürer, his art, and his ideas about geometry, the Metropolitan Museum of Art in New York City has some of his engravings on their site (www.metmuseum.org) including *Melencolia.* We have talked a lot about ancient Greece in our explorations to this point in the book because they discovered so many of the basic ideas. However, in the Renaissance art and engineering overlapped often (for example, with Leonardo da Vinci) and we explore some of that in Chapter 13.

ANSWERS
CONE

If you doubled the radius, what would that do to the surface area? The formula for the surface area of a cone is:

$$\text{Surface area (original radius)} = \pi r \sqrt{r^2 + h^2} + \pi r^2$$

If we were to double the radius, r (while leaving the height alone) we would increase the surface area as follows:

$$\text{Surface area (double radius)} = \pi 2r \sqrt{(2r)^2 + h^2} + \pi 4r^2$$

We would not quite increase the surface area by a factor of 4 since the term proportional to the height would only double. But, depending on the relative size of radius and height, we would be close to a factor of 4 if the height was small and farther away if the height was large relative to the radius.

CYLINDER

If you doubled the height, what would that do to the surface area?

$$\text{Surface area (original height)} = 2\pi r(h + 2r)$$

Doubling the height would give us

$$\text{Surface area (original height)} = 2\pi r(2h + 2r)$$

Depending on the relative size of h and r, it might nearly double the surface area (if the cylinder is tall and skinny and h is many times greater than r). If the height was small relative to the radius, then the surface area would be much less than double in size if h were doubled.

CHAPTER 10
SLICING
AND
SECTIONS

In Chapter 8, we learned how to figure out how much volume is contained inside different 3D shapes, and in Chapter 9, we computed the surface area of many of them. In this chapter, we explore what happens when we slice a prism, cylinder, or cone with an infinitely long plane. The resulting 2D shape is called a *section*. You are probably familiar with cross-sections in mechanical drawings of parts; those are typically cut parallel to an axis. We will explore what happens when we are more creative than that in lining up our cutting plane.

Slicing cones in particular gives surprising results. Our friends the Greeks (particularly the mathematician and astronomer Apollonius of Perga, who lived about 2100 years ago) figured out that slicing a cone in different ways created *conic sections*: circles, ellipses, parabolas, and hyperbolas. Which one you get depends on the axis of rotation of the plane of the slice. We'll give you a good start on the intuition you'll need for more advanced work later on. In Chapter 11, we'll learn about applications for conic sections, from the arc of a ball thrown up into the air to the orbits of the planets.

SLICING PRISMS

As we saw in Chapter 8, a prism is a 3D shape with identical, parallel polygons at both ends, and flat faces connecting the sides. A cube is a prism, too, which happens to have all its sides the same and square faces, base, and top. When we take a plane and use it to slice a prism in two, the 2D section that results is often counterintuitive. For example, depending on how you cut through a cube, you may get a cross-section that has anywhere from three to six sides.

MODIFYING THE MODEL

The model **cross-section.scad** can make slices from n-sided prisms (n = 4 being a cube). The model allows you to set the height of the cutting plane where it intersects the prism's central axis, and the angles through which the plane will be rotated — first about the x, then y, then z axes. It will be easier to think about this angle if you don't try to rotate the same cut in both x and y. Instead, raise the intersection point to where you want it to be, then tilt the plane about the x- or y-axis, then (if necessary) rotate in z to align it in the direction you want.

3D Printable Models Used in this Chapter

See Chapter 2 for directions on where and how to download these models.

cross-section.scad
This model creates an n-sided prism (or approximates a cylinder if n is about 100) and a plane that intersects it at variable heights and angles.

sphere_section.scad
This model cuts a sphere with a plane, resulting in a sphere cut into two pieces

conic_sections_set.scad
This model slices a cone to produce all the conic sections.

conic_section.scad
This model slices a cone with a user-specified plane, and prints both halves of the cone.

YOU WILL ALSO NEED:
- Flashlight (small and bright is best)
- A tube from inside a paper towel or toilet paper roll

FIGURE 10-1: The rectangular cross-section of a cube

The model creates a raised edge on the cut face on one side and a slightly depressed edge on the other. This will allow you to find the cut surface easily by touch and helps the models stay together a little better when you assemble them back into the original polygon. The model automatically rearranges the cut pieces to print as easily as possible. The **cross-section.scad** model has the following variables (shown with their default values):

- n = 4;
    - Number of sides of the prism. For a cube, n = 4. To approximate a cylinder, make n = 100.
- h = 50;
    - Height of the prism, in mm
- r = h / (2 * cos(180/n));
    - Radius of the prism, in mm (default makes a cube)
- sliceangle = [45, 0, 0];
    - The angle of the slicing plane, about the x, y, and z axes, in degrees. You should only rotate in x or y (not both), and this angle should be within +/-90 degrees to avoid unnecessary overhangs.
- sliceheight = h / 2;
    - Height along the side of the figure at which the cutting plane enters the prism. Use h / 2 if you want to cut through the prism's center.

SLICING CUBES

Depending on how we slice a cube with a plane, we will get different types of cross-sections. The number of sides the polygon will have depends on how many faces of the cube we cut through. Thus, there can be from three-sided to six-sided figures cut from a cube. These will not necessarily be regular

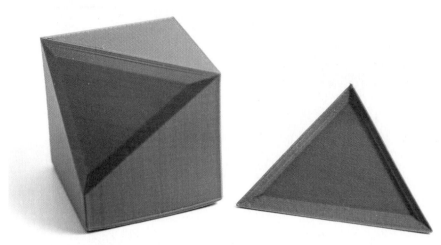

FIGURE 10-2: Triangular cross-section of a cube

polygons (where all sides are the same length) and some regular polygons are not possible when cutting various polyhedrons.

RECTANGULAR (4-SIDED) CROSS-SECTION

If we think about slicing a cube and are asked what shapes a plane would make cutting through it, the first thought would be a square (or rectangle if the cut is slanted). We can see a cut like that in Figure 10-1, made with the parameters:

```
sliceangle = [45, 0, 45] and
sliceheight = h / 2
```

This means we took our cutting plane and rotated it first about the x-axis 45°, at a height half the height of the cube above the plane of the base. Next, we rotated the plane 45° about the z-axis. The cut crosses four sides of the square, and the cut face is a 4-sided polygon (Figure 10-1). Note that it is also possible to have non-rectangular 4-sided cross-sections, depending on the angle of the plane.

TRIANGULAR (3-SIDED) CROSS-SECTION

Now let's think about how we might make a triangular cross-section. The only way we can get a straight side of a cut is to have the plane intersect a side. So, to get a three-sided figure, we want to cut off a vertex of the cube so we are slicing through three, and only three, sides. In the case shown in Figure 10-2, we used

```
sliceangle = [45, 0, 0] and
```

FIGURE 10-3: Pentagonal cross-section of a cube

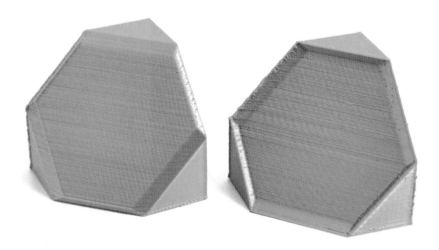

FIGURE 10-4: Hexagonal cross-section of a cube

```
sliceheight = 0
```

PENTAGONAL (5-SIDED) CROSS-SECTION

To get a 5-sided cross-section, our plane needs to intersect five of the six sides of the cube. In order to do that, we'll have to have the cut skim through the cube so that one side is left untouched. For the pentagonal cross-section in Figure 10-3, we used

```
sliceangle = [45, 0, 0] and
sliceheight = h / 4
```

HEXAGONAL (6-SIDED) CROSS-SECTION

To get a 6-sided cross-section, our plane needs to intersect all six sides of the cube. That means that the cut needs to go through the top, bottom, and all four sides. One way to do that is to imagine that you have the cube balanced on one of its vertices, with the plane parallel to the ground.

As you raise the plane above the ground, initially you will have a triangular cross-section. As you continue to raise the plane parallel to the base of the cube, though, eventually it will reach the point where it intersects all six sides. If you start with a cube sitting on one face on a table and rotate it 55 degrees around one of the axes in the plane of the table, you'll have it on one corner. Therefore, to get the hexagonal cross-section in Figure 10-4, we used:

```
sliceangle = [55, 0, 0] and
sliceheight = h / 2
```

COMPARING CASES

If we were to compare all these cases, assembling the cubes from the one with a triangular cross-section through the one with the hexagonal cross-section, what would we notice?

Most of the cross-sections will not be regular polygons (that is, shapes with all the sides the same). To get a regular polygon, the cutting plane would have to intersect each side of the cube for the same distance, which only happens for certain angles and axes of rotation of the cutting plane.

Examine the different polygons and make observations about the angles at which the sides come together. How would you orient the plane through a cube to make a square versus a rectangle? (Answer: a cube would require cutting through parallel to a face.)

SLICING OTHER PRISMS

A cube is a special case of a prism (a polygon with parallel top and base, and n sides). For a cube, n = 4. A 4-sided regular polygon is a square, and extruding it gives you a square prism. To make that square prism a cube though, you need to make the square's sides equal to the prism's height. Since we are using the cylinder primitive in OpenSCAD, and it requires us to specify a radius, we need to calculate the radius of the square.

FIGURE 10-5: 8-sided cross-section of prism

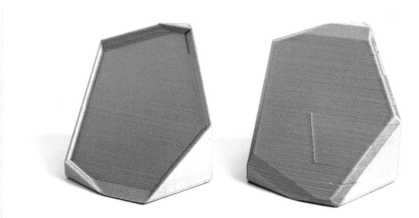

FIGURE 10-6: 7-sided cross section of prism

A regular polygon's radius is the distance from its center to each vertex, and its apothem is the minimum distance from its center to any side. The ratio between these two values for an n-sided regular polygon is cos(180° / n). Setting the radius to half of the prism's height divided by this ratio means that for any even number of sides, the width (measured from one flat side to another) will be equal to the height. (This is the default in **cross-sections. scad**.) See Chapter 7's discussion about inscribed and circumscribed polygons for the details.

If you think about it, an n-sided prism has n+2 sides. *This is also the maximum number of sides a polygonal cross-section can have*, for the same reason that a cube can have at most a hexagonal (6-sided) cross-section. Here are two examples, and the appropriate parameters for cross-section.scad. Figure 10-5 shows an 8-sided cross-section, from a 6-sided (hexagonal) prism. Figure 10-6 shows a 7-sided cut through a pentagonal prism. (A hexagonal prism has a top and bottom, so a total of 8 faces; a pentagonal prism has a total of 7 faces.)

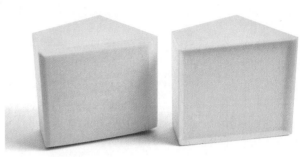

FIGURE 10-7: Rectangular-sided cross-section of prism

You might follow along with this by testing out the OpenSCAD parameters we give for each example, and seeing for yourself how the cross-sections change as you play with the given parameters.

FIGURE 10-8: Elliptical cross-section of a cylinder.

8-sided cross-section (Figure 10-5):
```
n = 6
sliceangle = [50, 0, 30]
sliceheight = h / 2
```

7-sided cross-section (**Figure 10-6**):
```
n = 5
sliceangle = [45, 0, 0]
sliceheight = h / 2
```

Finally, any right prism will have a rectangular cross-section if you slice it from top to bottom. In Figure 10-7 our example was a pentagonal prism, with a total of 7 sides including the top and bottom.
```
n = 5  (but any value will work)
sliceangle = [90, 0, 0]
sliceheight = h / 2
```

SLICING A CYLINDER

Imagine that we make prisms that have more and more sides. Each side gets shorter, and, as we saw in Chapter 7 for polygons and circles, the prism starts to approach a cylinder. Cutting straight across it (with the plane parallel to the base of the cylinder) will give us a circle. Cutting through it at an angle will give us an *ellipse*, which is like a circle that is stretched out

FIGURE 10-9: A sliced sphere

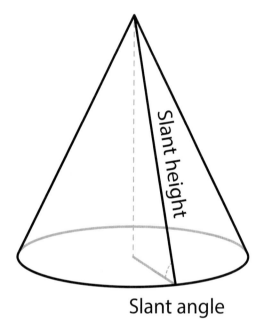

FIGURE 10-10: Slant angle and slant height

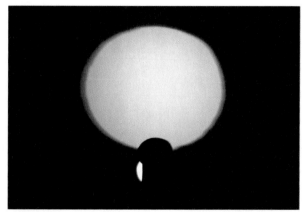

FIGURE 10-11: A flashlight making a circle on a wall (slightly distorted by tube and photo angle)

along the other axis. We will learn more about ellipses later in this chapter.

If the angle is too large for the cylinder's height, one or both ends of the ellipse will be chopped off where it intersects the ends. If we rotate all the way to 90 degrees, the sides will become straight, and we'll get a rectangle, just as we saw for the prism. To create the model in Figure 10-8, use these parameters:

```
n = 100
sliceangle = [30, 0, 0]
sliceheight = h / 2
```

SLICING A SPHERE

Imagine you have a sphere. If you cut it with a knife, what does the slice look like? No matter how you cut through a sphere, you will get a circular cross-section (Figure 10-9). All cuts through the center will be identical circles, and cuts that don't go through the center will also be circles, but smaller ones.

Since a sphere is perfectly symmetrical around its center, the angle of the cut doesn't matter. All that matters is the distance from the center of the sphere to the closest point on the cutting plane. You can see a sliced sphere in this photo from model **sphere_section.scad**.

The model has just two parameters: **size**, which is the sphere's diameter (in mm), and **sliceheight**, which is how far from the center of the sphere the cutting plane is intersecting it.

CONIC SECTIONS

When we cut prisms earlier in the chapter, we always wound up with another polyhedral cross-section. When we cut a cylinder, we got a circle or an ellipse (or a rectangle, if we cut through it perpendicular to the base). Cones are a bit more complicated, though. The *conic sections* resulting from slicing through a cone with a plane are circles and ellipses, along with two new shapes: parabolas and hyperbolas. (Hyperbolas require two cones arranged point-to-point, but we'll get to that later.) As we'll see, we can predict, based on the angle of the cut, which of the conic sections we will get.

First, we'll do a simple experiment with a flashlight. Then we will go through a 3D printed model of a set of all the conic sections, then experiment a bit with making one cut at a time.

Chapter 11 will explore ways to construct the conic sections, tied to situations where the conic sections come up in real life. Actually deriving the equations creeps a bit too far into the pre-calculus world for this book, but we will give you pointers to other resources if you want to go that far.

Speaking of math we need, in the rest of the chapter, we will use the Pythagorean Theorem and the trigonometric functions (sine, cosine, and tangent) a lot, so you might need to dip back into Chapter 6 a bit first for a refresher.

SLANT ANGLE AND SLANT HEIGHT

The *slant angle* of a cone is the angle the side of the cone makes with the base. If we have the height, h, of the cone, and the radius of the base, we know that the tangent of the slant angle has to be the height divided by the radius.

Slant angle = $\tan^{-1}$(height/radius)

Note that some books measure the angle of the cutting plane relative to the vertical axis of the cone, so be sure and know what convention a book is using if you read about conic sections. We will measure the angle from the base of the cone, with some y- or z-axis offsets of the cutting planes in some examples to make various points.

The *slant height* is the distance from the base of the cone to the vertex, along a side. It is the hypotenuse of the triangle made by the base and the height (**Figure 10-10**). From the Pythagorean Theorem,

Slant height = $\sqrt{r^2+h^2}$

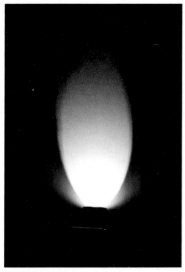

FIGURE 10-12: A flashlight making an ellipse on a wall

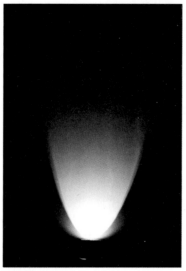

FIGURE 10-13: A flashlight making a parabola on a wall

FIGURE 10-14: A flashlight making a hyperbola on a wall

PRELUDE: FLASHLIGHT EXPERIMENT

If we take a flashlight and point it at a wall, we are creating a cone of light. Let's try tilting that cone relative to the wall to see if we can create the conic sections. If you have a flashlight smaller than a toilet paper roll and can shine the flashlight through the tube at the wall, the following experiment will have better results than just a flashlight with stray light diffusing everywhere. In our case, we took a cell phone (which has a good lens over its light) and put it in flashlight mode. Then we cut about a 2-inch long piece of a toilet paper tube to make the light come out nice and parallel, and held that against the phone. You can see the toilet paper tube in the images. You can use a regular flashlight, but they tend to have parabolic reflectors (more on that later) so you get more complicated effects. We will refer to a "flashlight" in what follows as meaning a light through a toilet paper tube.

Take the flashlight and shine it straight at the wall. You will get a circular spot of light on the wall (Figure 10-11). But if you start to tilt the flashlight a bit, the circle will stretch out along one axis and become an *ellipse* (Figure 10-12). If you tilt it at an even greater angle to the wall, eventually one end of the ellipse will open up, and you have a *parabola* (Figure 10-13). Finally, if you tilt the light even further, you will get one branch of a *hyperbola* (Figure 10-14). If you do an internet search on "conic sections flashlight" you will find video examples, too.

MODIFYING THE MODELS

Let's fine-tune our understanding of when a cut will give us one or the other conic sections. The model **conic_sections_set.scad** prints a cone with cuts that create the various cross-sections. You might need a bit of double-sticky tape or museum putty to hold this model together, depending on the slipperiness of the filament you are using. For the photos in this chapter, we printed two sets using different color filaments and re-assembled the cones while alternating the pieces to make the cuts more visible (Figure 10-15). As with the prism slice models, the two sides of the cut are beveled on one side and have a lip on the other. This makes it clearer where the cut is, and helps them hold together better when you assemble the set.

FIGURE 10-15: The set of all the sections (model printed twice)

The model **conic_sections_set.scad** has the radius of the cone and its height equal to each other, and we do not recommend changing any parameters since the relative arrangements of all the cuts are carefully calibrated not to cross each other. This model is best scaled in your slicer if it needs to be bigger or smaller. Be sure that you scale all axes by the same amount.

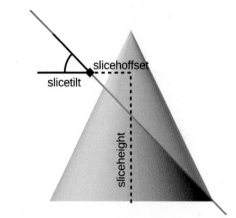

FIGURE 10-16: Definitions of sliceheight and slicehoffset

If, on the other hand, you want to explore just one cut through a cone, the model **conic_section.scad** has the following parameters you can change:

FIGURE 10-17: Both sides of the circular cross-section cut.

- **r**, the radius of the cone, in mm
- **h**, the height of the cone, in mm
- **slicetilt,** the angle (in degrees) about which the cutting plane is rotated
- Two parameters for the position of the axis about which the cutting plane is rotated: **sliceheight** in the z-direction (mm), **slicehoffset** in the plane of the base of the cone (mm).

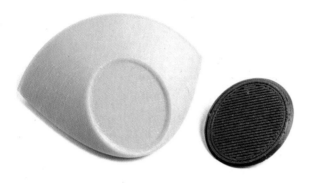

FIGURE 10-18: Both sides of the ellipse cut

FIGURE 10-19: Parabola slice when cone height and radius are equal

If **sliceheight** and **slicehoffset** are both set to zero, the axis of rotation of the plane is also the diameter of the base of the cone. Since this can be a little tough to visualize, we suggest going into OpenSCAD and playing with **slicetilt, sliceheight,** and **slicehoffset** to see how changing them affects the placement of the cutting plane (Figure 10-16).

If you don't have a 3D printer, you could consider creating a cone out of modeling clay and cutting it at an appropriate angle. You'll need a protractor to measure the slant height and the angle you are cutting the cone, and a way to cut it that won't distort it.

CIRCLE CROSS-SECTION

Now, let's deconstruct the model. If we cut through a cone with a plane parallel to its base, the cut we make is a circle. How big a circle it is depends on how far up from the bottom of the cone we make the cut. But any cut parallel to the base will be a circle (Figure 10-17).

Right at the vertex of the cone, the cut would be a point, but that is a special case. We have a short section later in the chapter about the more general issues that arise when any cutting plane passes through the cone vertex.

ELLIPSE CROSS-SECTION

If, however, we tilt the cut so that we are cutting at an angle to the horizontal that is more than zero, but less than the slant angle, we get an *ellipse* — a circle stretched out in one dimension (Figure 10-18). Ellipses come up all the time in astronomy; the orbits of the planets around the sun and the moon around the earth are ellipses. We will see how to derive an equation for an ellipse, and an alternate definition of it, as part of our applications of conic sections in Chapter 11.

The more we tilt the plane, the longer and skinnier the ellipse gets. Like the circular cross-sections, the ellipse will be larger if the cutting plane doesn't pass very close to the vertex. It's pretty obvious why we get an ellipse if the cutting plane is at an angle to the horizontal greater than zero, since cutting at a slant elongates one axis. It's less obvious why the slant angle is the upper bound. Let's talk about the parabola to see where that boundary comes from.

FIGURE 10-20: Parabola slice when cone height is 3 times the radius

PARABOLA CROSS-SECTION

What happens if we cut through the cone parallel to its side (that is, when the cutting plane is at the slant angle)? The resulting cross-section is a curve called a *parabola* (Figure 10-19). Parabolas make their appearance in many physical situations, notably in parabolic dish antennas or mirrors that concentrate light in one spot called a focus. In Chapter 11 we'll talk about why those systems work. For now, let's just see what they look like and how they arise from cutting a cone.

As we saw earlier in the chapter, the slant angle is:

Slant angle = tan⁻¹(height/radius)

In this case, since the radius of the cone and its height are the same,

Slant angle = tan⁻¹(1) = 45°. We can see the parabola in Figure 10-19.
On the other hand, if the height was 3 times the radius (Figure 10-20), we would get:
Slant angle = tan⁻¹(3) = 71.6° (relative to the base of the cone).

We used the model **conic_section.scad** with the following values to create two cones, one with the height equal to the base radius and one with the height three times the radius. Other parameters were the same, as shown. We then tilted the plane appropriately to get a parabola, as we just calculated.

FIGURE 10-21: Parallel slices, showing half of each model

FIGURE 10-22: Parallel slices, with the model assembled

Height = radius	Height = 3 * radius
h = 50;	h = 50;
r = h;	r = h/3;
sliceheight = 0;	sliceheight = 0;
slicehoffset =0;	slicehoffset =0;
slicetilt = 45;	slicetilt = 71.6;

Now let's look at what happens if we pivot the plane about a different axis of rotation when making the cut. We printed the cone with **h = r** and a 45° cut and got a parabola as expected. However, when we created the cone showing all the cuts at once, we needed to do a bit of adjusting so that the cuts would not cross each other. It turned out that the offsets for the parabola needed to be:

```
sliceheight = h * 3 / 4;
slicehoffset = r * 1 / 4;
```

The silver cone shown was printed with these parameters (the same as in the cone showing all the cuts). The bronze one was printed with both **sliceheight** and **sliceoffset** equal to zero. You can see that the slices are parallel to each other, but result in a differently scaled parabola (Figures 10-21 and 10-22).

Unlike an ellipse or a circle, the parabola is not a closed curve. The straight part of the cross-section along the bottom of the cone is not considered part of the parabola. Planes cutting at the slant angle (and more steeply) create curves that are not closed (parabola and hyperbola.) You can see why if you look at the parabola's cross-section: any cut steeper than that would always be cut off at the bottom of the cone. If the cone was infinite, the sides of the cut would go on forever.

FIGURE 10-23: Hyperbola cross-section across the pair of cones at 90 degrees

HYPERBOLA CROSS-SECTION

If we cut through the cone at an angle greater than the slant angle, the cross-section is a *hyperbola*. To see the entirety of a hyperbola you need two cones, placed vertex to vertex. A hyperbola is a pair of curves (often called branches) that don't touch each other but go off to infinity in opposite directions.

As we noted earlier, the ideal versions of these curves are defined on cones that are infinitely tall, and so the hyperbola's branches would go off to infinity. In fact, as we'll see in Chapter 11, if a comet or spacecraft comes screaming in toward the sun too fast to get into an orbit around it, that's a hyperbolic trajectory. Hyperbolas come up in optics quite often too, merged with other conics to get rid of effects like the rainbow halos called chromatic aberration that otherwise appear around objects in a telescope.

The simplest way to create a hyperbola is to cut a cone at 90° to its base. Any such cut that does not go through the vertex will result in a hyperbola. This pair of cones was printed with the height (**h**) set to 70mm, and the radius (**r**) set to **h/2**. The cut was halfway between the center of the cone and its edge, which used these parameters in **conic_section.scad**.

As you can see, the cuts are symmetrical about the axes (Figure 10-23).

FIGURE 10-24: Asymmetrical hyperbola cross-section

Model parameter	Top cone	Bottom cone
h	70	70
r	h/2	h/2
slicetilt	90	90
sliceheight	0	0
slicehoffset	r/2	r/2

It gets a little trickier if we want to try out making a hyperbola with a plane that cuts at an angle other than 90°. Then, the cutting plane will intersect the bottom and top cone at the same angle, but create a different cross-section in each cone and exit along a different line through their respective bases.

In the example we will go through here (and that is 3D printed with red translucent filament), the angle of the cutting plane is 80°. For that example, once again the height (**h**) is 70 mm, and the radius (**r**) is **h/2**.

On the lower cone, we used an offset of r/8 to the left of the centerline for the axis of rotation on the cutting plane but kept it at the height of the bottom of the cone, or **sliceoffset** = **0**. The centerline of the cone is an offset of 0, so this worked out to make the variable **slicehoffset** = **-r/8**.

So we have our plane of rotation rotating about a line in the plane of the base of the lower cone, and **r/8** offset to the left of center. If we want to define the same plane from the point of view of the top cone (in other words, drawing the cutting plane and looking at the pair upside down) this same cutting plane will look like a plane at height **2*h**, and offset **r/8** to the *right* (and thus of the same value, but the opposite sign) as the offset of the lower cone.

The red cones in Figure 10-24 illustrate this example. You can observe that the hyperbola intersections are a little different from each other. The branches of a hyperbola will be symmetrical in an infinitely-long pair of cones, but since the cones here are finite we see slices that are a little different from each other in the two cones. The parameters used for the two cones are summarized in the tables that follow.

Model parameter	Bottom cone	Top cone
h	70	70
r	h/2	h/2
slicetilt	80	80
sliceheight	0	2*h
slicehoffset	-r/8	r/8

If you want to try creating a different version of these cones, you may find it helpful to draw the diagram we drew above, and work out the values. Also, remember that **slicetilt** has to be greater than the slant angle to produce a hyperbola. It's very easy to mess up the signs of the offsets, so you may want to figure them out by walking through the geometry in detail as we did. For many values of **slicetilt** for a given cone, there will be no intersection between the plane and one of the cones, since the intersection would take place farther down the (hypothetical) infinitely-long cone.

CONIC SECTIONS AND ANGLES

In summary, then, we can predict what cross-section we will get by making a slice across the cone by knowing the angle of the cutting plane, as shown in this table.

Section	The angle of the cutting plane (not through the vertex of the cone)
Circle	0° (parallel to the base of the cone)
Ellipse	More than 0°, less than slant angle
Parabola	Equal to slant angle
Hyperbola	Greater than the slant angle, up through a maximum of 90°

SLICES INTERSECTING THE VERTEX OF A CONE

There are special cases if your plane cuts through the vertex (top point) of the cone. If your cut is between 0° and the slant angle (relative to the base), you will get a point. If the cut is at the slant angle (and would give you a parabola anywhere else), you will get a single line along the slant height of the cone. Finally, if you cut through the vertex more steeply than the slant angle, you will get two lines that meet at the vertex (crossing there for a double cone).

SUMMARY AND WHERE TO LEARN MORE

In this chapter, we have done an extensive walk through many ways to slice up prisms, spheres, pyramids, cylinders, and, finally, cones. We spent some time getting general intuition first about what type of cut through a cone would generate each of the conic sections (circle, ellipse, parabola, and hyperbola).

Good general references on the topic in this chapter, particularly conic sections and how they relate to one another can be found in the Wikipedia article *Conic Section*, and in many videos on the Khan Academy's site (khanacademy.org). The YouTube video mathematician 3Blue1Brown has many related videos on his channel, most notably *Why Slicing a Cone Gives You an Ellipse*. Finally, Paul Lockhart's book, *Measurement* (2012, Belknap/Harvard University Press) takes a similar hands-on approach to the one we make here, with more historical background.

In Chapter 12 we will learn about some properties of the conic sections and, broadly, where those equations came from. We will also see how common conic sections are in astronomy and physics applications.

CHAPTER 11
CONSTRUCTING CONICS

Conic sections (which we introduced in Chapter 10) turn up in many applications. They come up all the time in phenomena as different as the path of a ball tossed up in the air, the orbits of the planets, and the designs of telescopes and satellite dishes. We will explore ways to draw them and suggest places you can find them in the wild.

The approach in this chapter is a little unconventional. Typically, people start with the same definitions of the relationships of various features of conic sections, and then (using a lot of algebra) derive equations from them. This gets messy pretty quickly and walks into algebra that is beyond what we assume you know how to do at this point.

Instead, we're going to use a very simple 3D printed model to construct the conics geometrically. The model is so simple that you could probably figure out alternatives with stuff you have lying about (more on this later), but the 3D printed model is convenient.

ELLIPSES AND CIRCLES

How are ellipses and circles different from each other? We can define a circle as a constant distance from one point. You learned that when you drew circles with a compass in Chapter 4. If you sweep the pencil point all the way around, you get a circle.

What about an ellipse, though? We saw in Chapter 10 that if we take a slice of a cone parallel to the base of the cone, the cross-section is a circle. However, if we start to tilt that cut a little, the circle will stretch out into an ellipse. It turns out that ellipses have some interesting properties, and we will start with the classic way to draw one.

Take a piece of cardboard or foam core, thick enough so that push pins won't poke all the way through. (We used two layers of a shipping cardboard box). Tape a piece of paper over this base. Take two push pins and use them to pin down the ends of a piece of string (Figure 11-1). You can tie a knot in the string to prevent it from unraveling and make a more solid anchor for the push pins. The string should be loose. About 25% longer than the distance between the pins should be about right. String used to tie up packages works, or you can use thinner string if you prefer.

3D Printable Models Used in this Chapter

See Chapter 2 for directions on where and how to download these models.

centerfinders.scad
A series of nested half-circular pieces used in the constructions in this chapter.

Other things you will need:
- Piece of foam core board or heavy cardboard that will hold a pushpin
- Some paper to draw on
- Two push pins or thumbtacks
- String (like package string) that is not stretchy
- A pencil
- A drawing compass

FIGURE 11-1: Setting up the string to draw an ellipse

FIGURE 11-2: Placing your pencil

FIGURE 11-3: Drawing the bottom of the ellipse

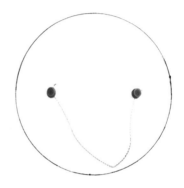

FIGURE 11-4: Finished ellipse

Now, put your pencil alongside the string, and stretch it taut, but not so tight that the push pins want to pop out (Figure 11-2).

Now, keeping the string taut, you can move the pencil around the push pins. You should be able to draw the top or bottom half of an ellipse(depending on where you started), and then adjust your pencil to make the other half (Figure 11-3).

Finally, you will have the entire ellipse (Figure 11-4).

Try moving the push pins farther apart. The ellipse will get longer and skinnier (Figure 11-05). On the other hand, if you imagined putting them right on top of one another (or, more likely, using only one push pin), you'd just have a circle.

The longer dimension of the ellipse (all the way from one end to the other, like a circle's diameter) is called the *major axis*. Half of it, like the radius of

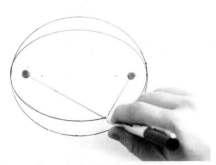

FIGURE 11-5: Moving the push pins farther apart

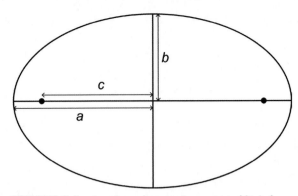

FIGURE 11-6: Semi-major and semi-minor axes and foci of an ellipse.

a circle, is called the *semimajor axis*, and usually referred to as "a". The narrower dimension is called the *minor axis*, and half of it is called the *semiminor axis*, usually shown as "b".

The places where our push pins reside are called the *foci* (pronounced "foe-sigh" or "foe-kigh", depending on where you are from). The singular is *focus*. The foci are a distance from the center of the ellipse that is often called "c" (Figure 11-6).

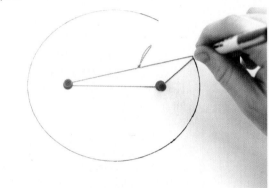

FIGURE 11-7: The loop of string variation

The dimensions a, b, and c of an ellipse are all related like this:

$$c^2 = a^2 - b^2$$

Where does that come from? Try to figure it out, then check out "Finding the locations of the Foci" in the Answers at the end of the chapter. (Hint: remember Pythagoras back in Chapter 6.)

You can also use a loop of string, instead of pinning the string (Figure 11-7). This has the advantage that it lets you run your pencil all the way around the ellipse without picking it up, but the math is just a bit more complicated. In this case, the circumference of your loop would be 2(a + c). If you want to see how the ellipse changes when you adjust the distance between the foci without changing the major axis, you would need to adjust the loop so that its circumference remains 2(a + c).

FOCI OF AN ELLIPSE AND INTERNAL REFLECTION
Our string was a constant length as we drew our ellipse around the circle,

held down at the foci at each end. That means that for any point on the ellipse, the sum of the distances from the two foci will be the same. This gives an ellipse some interesting properties.

Imagine that there was a light at one focus and that the inside of the ellipse was a mirror. Light that reflects off of a mirror will leave the surface at the same angle that it came from but in the opposite direction. Think of the string as a ray of light coming from one focus and reflecting back to another. No matter where the light from the first focus hits the side of the ellipse, it will reflect back to the second focus and will have to travel the same total distance. (A physicist would say that the light waves stay *in phase*). This is true for sound waves, too. Note that we are just talking about a two-dimensional abstraction here, and the behavior can get quite complicated in a 3D space.

WHISPERING GALLERIES

One feature of elliptical (and other curved) spaces is that sound from one focus can be heard particularly well at the other focus. This has been exploited in "whispering galleries." If you stand at certain places in some buildings with curved walls and domes you can hear sound from the other focus of the ellipse, even if it is far away and just a whisper. Most famously, this is said to occur at St. Paul's Cathedral in London and one place in Grand Central Terminal in New York City near the Oyster Bar. Whispering galleries occur in other curved rooms, and their physics is a little tricky. We'll talk about parabolic "acoustic mirrors" in the next section.

Sticking to ellipses, the sound from anywhere in this elliptical space (not at the foci) will arrive at the two foci as well, but will not be, well, focused in the same way that sound from the other focus will be. A physicist would say that they would *interfere* with each other, which just means that different sound waves bouncing around in the room would cancel each other out. However, a sound coming from the other focus will be clear even if there is a lot of other noise bouncing around the room.

It would be pretty tricky to make yourself a whispering gallery. The wall has to be hard and smoothly curved, which isn't something one encounters routinely. But if you are someplace with curved walls, you might see if you can find a place where this phenomenon will happen. Cloth, unfortunately, is likely to absorb sound, so hanging a sheet in a curve won't make you a whispering gallery. However, someplace with a lot of curved concrete (like a skate park) might have some interesting sound reflections.

CIRCUMFERENCE AND AREA OF AN ELLIPSE

When you draw an ellipse (or use the 3D printed model) you can also find the circumference, the distance all the way around it. (Sometimes people use the word *perimeter* for the distance all the way around anything but a circle, so you may see that too.) Lay down a piece of string around the ellipse, mark off the circumference on the string, and then straighten out the string on a ruler to measure the circumference. Bizarrely, there is no simple equation for the circumference of an ellipse, for reasons that need to wait until you've had calculus. Around 1914, the Indian mathematician Srinivasa Ramanujan came up with this approximation (a squiggly equals sign means "approximately"):

$$\text{Circumference of an ellipse} \approx \pi\left[3(a+b) - \sqrt{(3a+b)(a+3b)}\right]$$

It is a tribute to how hard this problem is that this approximation is only about 100 years old. Ramanujan, who lived in British India during its colonial period, did not have any formal training in higher mathematics but came up with this and other innovations largely on his own. He also developed better, more complex approximations as well; search on his name and "ellipse circumference" (sometimes called "ellipse perimeter") for more. Unfortunately, he died at age 32, or who knows how many other simplifications he would have come up with. The full solution involves an advanced calculus equation called an "elliptic integral", but this should serve you well for now.

The area of an ellipse is more straightforward. It is:

$$\text{Area} = \pi ab$$

To demonstrate it to yourself, consider that as an ellipse gets more and more like a circle, both a and b become a radius. So πab would become the more familiar πr^2.

ORBITS OF PLANETS AND KEPLER'S LAWS

For thousands of years, people believed that all the planets revolved around the Earth. They had to make up all sorts of weird explanations for the fact that planets seem to sometimes move one way on the sky, and sometimes another. In 1543, the Polish mathematician Nicholas Copernicus explained that the planets moved in circles around the sun. In the early 1600s German mathematician Johannes Kepler figured out that if he modeled the planets' orbits as ellipses with the sun at one focus rather than as circles, the Copernican model was even better at predicting the motion of the planets in the sky.

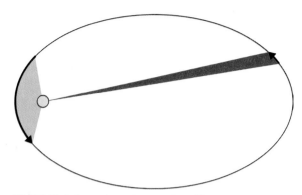

FIGURE 11-8: Sweeping out areas of an ellipse

FIGURE 11-9: The orbit of Halley's Comet

FIGURE 11-10: Orbits of Earth, Venus, and Mercury (mutually to scale, but not to the same scale as Figure 11-9).

Kepler organized his theory of how planets moved into three laws:

1.   Planets orbit the sun in an elliptical orbit, with the sun at one focus.
2.   A line dragged from the sun to a planet would sweep out an equal area of its orbit during an equal time (Figure 11-8). By "sweeping out", we mean if you imagine a string from the planet to the sun, as the planet moved around the cord would stretch (or contract) and drag along the imaginary plane of the orbit, creating the colored areas in Figure 11-8.
3.   The square of the period of an orbit (the time it takes a planet to go around once) is proportional to the cube of the semimajor axis of its orbit.

His laws, based largely on many careful observations by the astronomer Tycho Brahe, explained a lot. For example, planets closer to the sun, by the second law, have to move faster to sweep out the same area as a planet with a larger orbital radius.

Kepler didn't know about moons around planets, or planets around other stars. (Later astronomers would need to generalize his laws to allow for the gravitational pull of the bodies orbiting and some other factors.) Soon after, though, Galileo saw the moons of Jupiter through a telescope and corresponded with Kepler about it. It took a while, but eventually this view of how the solar system worked was accepted.

Figures 11-9 and 11-10 show a set of 3D printed models of Kepler's laws described in our book, *3D Printed Science Projects* (Apress, 2016). The base is an orbit, with the foci marked, and the height is the speed at that point in the orbit. Figure 11-9 shows the orbit of Halley's Comet, which has an orbit that is a long, skinny ellipse with the sun at a focus way out near the end of the semimajor axis. In the 3D print, the foci are pretty much buried in the wall at each end of the ellipse. Figure 11-10 shows the orbits and orbital velocities of Mercury, Venus, and Earth (to scale with each other, but not to Halley's Comet).

As you can see from the height of the model, the comet is moving very fast near the sun, then slows to a crawl at the other extreme as it moves along in its 75-year-long orbit. Mercury, too, has a somewhat elliptical orbit and will go faster in the part of its orbit nearer the sun than it will in the other. Looking at the set of three orbits, as we go farther from the sun, the orbits of Venus and Earth are each slower than the planets closer to the sun.

This variation of the speed of a planet, like Earth, in its trip around the sun is a major component in the Equation of Time. We learned about the Equation of Time in Chapter 7 when we had to look up its value to deduce our location based on when the sun was highest in the sky. Earth's orbit is very nearly circular, but even that small difference matters. In Chapter 12, we will reprise that, and go into more depth about how to take these effects into account when using the sun to tell time.

PARABOLA

As we saw in Chapter 10, a parabola is created only when a cutting plane is exactly parallel to the slant angle of the cone. If the cutting plane is any shallower (relative to the bottom of the cone) you get an ellipse; any steeper, you get a hyperbola.

Directrix and Generatrix

Getting further into the details of conic sections requires that we make the acquaintance of a new pair of abstract ideas, the directrix and generatrix (sometimes called a generator instead). Despite sounding like pets of superheroes, they are imaginary tools that can help us draw a curve. Since we are all about being hands-on, we are going to make and use physical versions in the next sections of this chapter.

A generatrix is a point, curve, or surface that when moved along a certain path carves out a desired line, surface, or 3D shape, respectively. A directrix is that path.

For example, take a straight line that has the top end pinned at a point and the bottom swept around a circle to sweep out a cone. The line is the generatrix, and the circle is the directrix. We are going to use a directrix to construct conic sections (specifically, hyperbolas and parabolas) without (much) algebra, in case you want to make something physical in one of these shapes. You know how to make an ellipse now (with a pinned string or rope). Let's see how to draw parabolas and hyperbolas, and why you might want to.

A parabola can be thought of as the special case where an ellipse transitions into a hyperbola. It is shaped somewhere between the letters U and V (never quite coming to an actual point). The sharpest point of its turnaround is called the *vertex*. Some people like to think of a hyperbola as an inside-out ellipse. After we talk about all three, we'll come back to that.

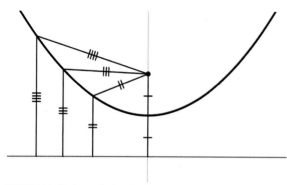

FIGURE 11-11: A parabola, showing its directrix line (red), focus, and lines of equal distance

FIGURE 11-12: Finding the vertex of the parabola (lower dot) from focus (upper dot) and directrix (line)

After reading the last section, you probably won't be surprised to know that a parabola can be defined in terms of a focus, but only one. An alternative way to think about it is to imagine the ellipse we get from slicing a cone. As the cut gets steeper, the ellipse gets larger, and the major axis gets larger faster than the minor axis. When the cut is parallel to the side of the cone, the ellipse would never cross the side of the cone to close the ellipse. We can imagine that one focus has slid out to infinity. Our trick of using two foci we used to draw the ellipse won't work here, but we can do something almost as simple.

DRAWING A PARABOLA

In addition to its conic section definition, a parabola can also be described as a curve that has every point an equal distance from a point (the focus) and the shortest distance to a line (the directrix). In Figure 11-11, we show these distances marked with little hash marks. Lines with one hash mark are equal in length to each other, the ones with two are equal to each other, and so on. The directrix is shown in red.

In the "Directrix and Generatrix" sidebar we get into this a little more, but, briefly, once you pick your directrix and focus, your parabola has been completely specified. Just like cutting a cone parallel to its side, picking a directrix and focus is a means of generating a parabola geometrically. A parabola can open at any angle. The parabola will always open away from the directrix, with the focus on its center-line. If the directrix is above a focus, then the parabola will open downward, and if the directrix is below the focus, the parabola will open upward.

The lowest point of the parabola in Figure 11-11 is called its *vertex*. It lies along a line (shown in blue in Figure 11-11) that is perpendicular to the directrix and goes through the focus. The focus and directrix are equal distances from the vertex, and this is the closest the curve approaches to either.

We've created a 3D printed model, **centerfinders.scad,** which allows you to play with these concepts. We can use it to create both hyperbolas and parabolas. It's the equivalent of what we did with the push pins and string for an ellipse. Here is how it works.

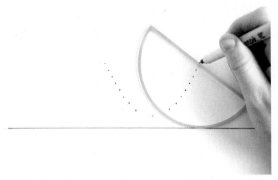

FIGURE 11-13: Starting the parabola

FIGURE 11-14: The parabola has two symmetrical points on either side of its vertex.

If we want to draw a parabola, we have to find a series of points that are the *same distance from the focus and the directrix line*. One way to do that is to draw a series of circles that each just touch the focus and the directrix, since a circle centered on one point of the parabola will have the focus and directrix both touch its perimeter somewhere. However, it's tough to draw a circle if you don't know where the center is (since that is the point we are trying to find).

The **centerfinders.scad** model prints out a series of shapes that look like a capital letter "D". To draw a parabola, first draw a directrix and a focus. Find the spot midway between the focus and the directrix with a ruler; that will be the vertex of the parabola (Figure 11-12).

Then take the smallest D and just touch the directrix and focus with the outer part (Figure 11-13). Make a dot in the notch on the flat point. That is the next point of the parabola.

Then, take the next-biggest D and so on. Note that each D can make two points - one on each side of the vertex, since a parabola is symmetrical (Figure 11-14).

Once you have this series of points, you can sketch in the final parabola between them (Figure 11-15).

If you want to adapt **centerfinders.scad** a bit, you can change how many D-shaped pieces you make, how large they are, and how different the size is from one to the next by changing the radii set of three variables.

The default is: `radii = [10:6:60];`

FIGURE 11-15: The final parabola

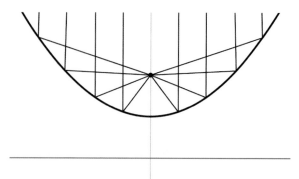

FIGURE 11-16: Light coming into a parabola

This, as we saw in Chapter 2, means that the smallest radius piece will have a radius of 10mm, and we step up 6mm at a time until we reach at most 60mm (here it would be 58). Here is what the other variables do, shown as in-line comments. All measurements are in millimeters.

```
w = 2; //line width of each "D"
h = 3; //the thickness in the vertical direction
hole = 2; //radius of the notch to draw the parabola
point
```

The variable **enclosed** set equal to **TRUE** will draw a whole circle around the place you will draw the parabola point. Otherwise, it creates a half-circle notch. If you change its value to **TRUE** you might need to increase the spacing (second value in the **radii** set) to keep them from overlapping.

If you don't have a 3D printer, you can set **h = 0**. Render the file as usual, but then use **File > Export > Export as SVG**. You can then drag an SVG file into a browser and paper-print it, or prepare it to be laser-cut. There is a tool called a "round center finder compass" normally used for wood carving that might be useful, too.

PARABOLIC MIRRORS AND ANTENNAS

It's cool to be able to draw a parabola, but why do we care about parabolas, and what physical properties are implied by the construction we just did? Let's imagine that we have light streaming into a parabola that is mirrored on the inside. The light waves can be thought of as all coming in perpendicular to the directrix(red line in Figure 11-16).

When they hit the parabola, each light wave will be reflected at the same angle that it came into the parabola, just as on an ellipse. That means that all the light entering the parabola will be concentrated into the focus.

Just as sound in an elliptical space will come to the foci, beams of light coming into the open end of a parabola such that they are perpendicular to the directrix will all bounce off the parabola and wind up at the focus. Alternatively, a light at the focus with a parabolic mirror behind it will result in parallel beams all emerging in a direction perpendicular to the directrix. This is why satellite dishes and other devices gathering radio or light waves often take the shape of a parabolic dish with a receiver (or transmitter) at the focus.

Flashlight reflectors are another example of parabolic reflectors. Some light will be cast directly from the source (bulb/LED) and will disperse quickly, but many flashlights also have a reflector that, to varying degrees, approximates a parabola. This is why you often see a brighter but narrower inner beam. The more closely the reflector approximates a parabola (and the closer the light source is to its focus), the narrower the beam will be, and the less fall-off in intensity you will get with distance. Special long-throw flashlights are designed with highly parabolic reflectors and carefully-placed sources, so that light rays will be very parallel, and you can brightly light a point far away, but those beams will never illuminate a very wide area.

There are whispering gallery effects like the ones we saw with ellipses from strategically-placed parabolic walls, too. Sometimes called *acoustic mirrors*, curved walls or dishes can be used to collect and focus sound. Giant concrete acoustic mirrors were put in place along the south coast of Britain in World War II to detect the sound of enemy bombers before radar came along and made them obsolete.

KICKING A BALL INTO THE AIR

If you kick a ball into the air, it will fall on an arc that describes a parabola. On the way up to the high point, the force of gravity slowly decelerates your ball. (Air resistance plays a part, too, but we will ignore that.) Then, on the downward part of the trajectory, it will accelerate (Figure 11-17). The width of the parabola is determined by how much energy you impart in the direction parallel to the ground. How high it goes depends on the energy you impart in the upward direction.

You'll find out in calculus (or AP physics) that the time, t, it takes a ball to fall from its highest point, which we will call, h, is:

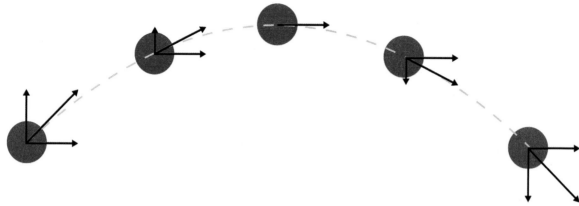

FIGURE 11-17: The kicked ball

$h = (1/2)gt^2$, which we can rearrange as:

$t = \sqrt{2h/g}$

Where **g** is the acceleration due to gravity, 9.8m/s² (on Earth, anyway). What that means is that for every second the ball is in flight, gravity drags it down at a speed that increases by 9.8m/s *every second*. Note that t = 0 is the time at which the ball is at its highest point, not when we kick it. So if your ball topped out at a respectable 10m, it would hit the ground in about

$\sqrt{2 * 10/9.8}$ = 1.4 seconds after reaching its highest point.

Note that this would be just half the time it spent in the air (and it might roll or bounce on the ground after that, of course). On the moon, the acceleration due to gravity is just 1.6m/s², so our ball would take about 3.5 seconds to fall 10 meters.

If you throw a ball straight up, it will come straight down rather than making a parabola. One of Isaac Newton's big insights when he came to understand gravity and forces on objects was that you could treat them as independent. Gravity affects the vertical part of your ball's trajectory, and the force of your throwing the ball affects the horizontal part of the flight. We show that with the arrows in Figure 11-17. The diagonal arrow is how much the ball is moving if you add up the components of its velocity in the horizontal and vertical directions.

Experiment and see how different amounts of force and the angle at which you release the ball will create different parabolas. You can look up "parabolic motion" or "ballistics" to learn more. Before Newton, medieval military engineers would calculate tables of what angle to aim their catapults to land on the defending fortress wall. Ottoman Turks were particularly good at this, and it made them much-feared siege attackers. Leonardo da Vinci was very into ballistics analysis too. It was all experimentally derived in those pre-Newton days since they had not figured out the physics behind it exactly.

HYPERBOLA

The hyperbola is the final conic section. You may suspect that a hyperbola has a focus or two, and indeed they do have a pair. If you recall from Chapter 10, a hyperbola has two "branches" that never touch. When we see them as a slice through a cone, only half will be on one cone. The other branch will exist on another cone point-to-point with the first. There are a few different kinds of hyperbolas, but they share common traits. Some people like to think of a hyperbola as an ellipse popped inside-out, with the foci on the outside of the curve.

A hyperbola is defined as the set of points such that the *difference between* the distances from the hyperbola to the two foci stays constant. Phew, that's a little hard to visualize. Fortunately, though, there is a clever way to construct one using the foci and a pair of *circular directrixes*.

CONSTRUCTING A HYPERBOLA

In Figure 11-18, we have drawn the two branches of a hyperbola that happen to open to the right and left. The foci are at points equidistant from the centerline. Imagine we draw a circle around the left focus of radius 2a. It will now be true that, for any point on the right-hand branch of this hyperbola, that the distance from the right focus to the hyperbola will be equal in length to the shortest distance from the hyperbola to that circle.

The diameter of that circle, plus the location of the foci, will determine the shape of the hyperbola. (We have shown these equal-distance lines by having equal numbers of hash lines on equal length lines.) Pretty similar to how the parabola directrix worked, isn't it? If we then drew a circle of the same diameter around the other focus, we could draw the other branch.

Thus we can draw a hyperbola with two foci and a pair of *circular directrixes*. We can use our same set of D-shaped pieces from **centerfinder.scad** to construct a hyperbola. First, draw a circular directrix (using a compass) and the two foci. The circular directrix needs to be centered on one focus and its radius needs to be such that the other focus is outside it. Then measure your

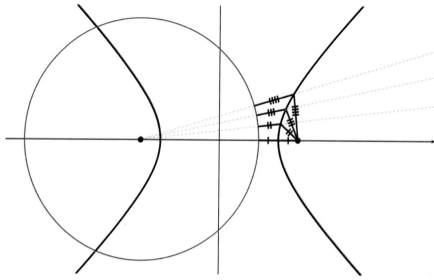

FIGURE 11-18: Hyperbola and circular directrix

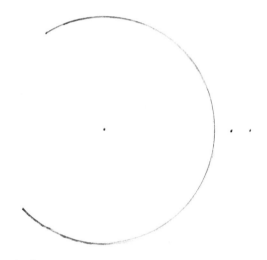

FIGURE 11-19: Marking the directrix, foci, and vertex of the hyperbola (the vertex is the center of the three dots).

first point, halfway between the circle and the focus outside the circle (Figure 11-19). That point will be the vertex of the hyperbola (as with a parabola, the vertex is the extreme point of the curve).

Next, use the D-shaped pieces to find the points that are an equal distance from the focus outside the hyperbola and the circular directrix (Figure 11-20 and Figure 11-21). Figure 11-22 shows the result after making a series of these marks. Finally, in Figure 11-23, we sketch in the hyperbola.

FIGURE 11-20: Marking subsequent points on the hyperbola

FIGURE 11-21: Farther along in drawing the hyperbola (note bigger "D")

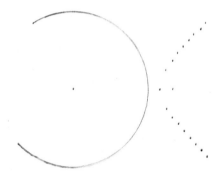

FIGURE 11-22: After marking a series of points

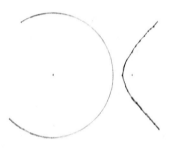

FIGURE 11-23: Sketching in the hyperbola

Finally, you can draw the other branch of the hyperbola by drawing a circular directrix around the other focus (of the same radius as the first one) and repeating the procedure (Figure 11-24). Eventually, the two branches of the hyperbola each start to approach a straight line, called the *asymptote*. There are two asymptotes, making sort of an x through the center between the two branches. As we get farther from the center, the hyperbola gets closer and closer to these lines. Their curvature also decreases, so that they approximate straight lines.

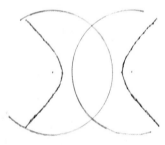

FIGURE 11-24: Drawing the other branch.

You can see that if the radius of the directrix changes, you will get a different hyperbola. The

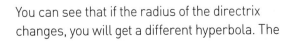

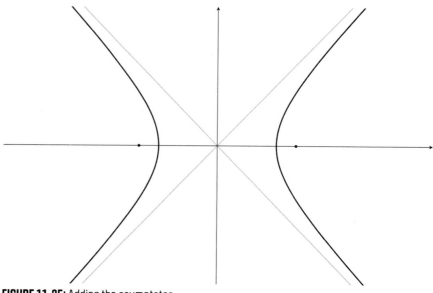

FIGURE 11-25: Adding the asymptotes.

angles at which the asymptotes cross will change, too (Figure 11-25). Try it!

HYPERBOLA APPLICATIONS

Not every orbit around the sun is a nice, tidy ellipse. Some objects (like comets) might come in from someplace way out in deep space at such high speed that they are on a hyperbolic trajectory. They may get close to the sun (or a planet, in the case of a spacecraft) but their relative velocity is so high that they will never settle into an elliptical orbit.

Sophisticated telescopes also often combine different types of mirror surfaces, like parabolas and hyperbolas, to eliminate distortions. If you were to make a hyperbolic reflector, light or sound waves generated at the focus would bounce off at the angles shown by the dotted lines in Figure 11-18.

TYING THE CURVES TOGETHER

Just as we could come up with all the conic sections by cutting a cone (or two) at steadily increasing angles, so too we can use the methods here to think of the conics as progressing from ellipse to parabola to hyperbola.

You can think of a hyperbola as an inside-out ellipse, with a parabola as the intermediate form where one flips to the other. Imagine starting with an ellipse. If you stay close to one focus and move the other focus away, the ellipse will get longer and thinner. To do this, it helps to imagine that you are allowing the ellipse to get larger as it stretches, so that the distance from the focus to the edge (that is, the distance a - c) remains constant. When the other focus is infinitely far away, your curve becomes a parabola. The lines going from any point on the edge to the distant focus will be parallel, because converging on a point infinitely far away is the same as not converging at all.

If you think about doing the same with a hyperbola, moving one focus further and further away while keeping the distance between the near focus and the curve's corresponding vertex constant, the other branch of the hyperbola will move off to infinity along with the distant focus. This time though, we'll choose the focus on the opposite side to be our distant focus. As the hyperbola stretches and gets bigger, the angle between its asymptotes gets smaller, and the point where they cross (which is halfway to the distant focus) also gets further away. When that distance is infinite, the asymptotes become parallel, and its curvature turns into that of a parabola.

Now, imagine these two parabolas are the same parabola. If you start with a circle (both foci in the same place), and move one focus out to infinity, it turns from a circle, to an ellipse, and then into a parabola, which behaves both as an ellipse with its distant focus infinitely far away in one direction, and a hyperbola with its distant focus infinitely far away in the other direction. Because the parabola can be thought of as both, you can think of the distant focus simultaneously being at both infinities, and able to move away toward one infinity then come back from the other, like Pacman going out of one side of the screen and coming back on the other side.

This also amounts to rotating the plane cutting our double cone in Chapter 10, as we progressed from circle, to ellipse, to parabola, and then hyperbola. There are some tricky ways to mark where the foci are when you are cutting a plane. If you take a sphere inside the cone that is tangent to the inside of the cone and the cutting plane, it touches the plane at the focus of the conic section. These spheres are called *Dandelin spheres*, and are more than we want to explore here, but you may enjoy looking them up to learn more.

SUMMARY AND LEARNING MORE

In this chapter, we learned about a different way of constructing conic sections. Later on, you can use the relationships in this chapter to develop equations for these three conic sections and explore them that way. You can find those derivations in a pre-calculus book, or by looking up "derivation of equation of an ellipse" (or parabola, or hyperbola).

We also learned about some places where ellipses, parabolas, and hyperbolas are used in real-life designs. They also come up when explaining the orbits of planets. In the next chapter, we'll learn a bit more about how our planet goes around the sun, and some ways to expand on our ideas in Chapter 7 about latitude and longitude.

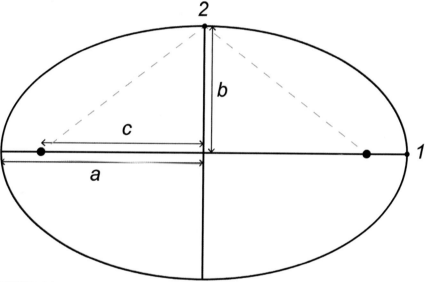

FIGURE 11-26: An ellipse.

ANSWERS
FINDING THE LOCATION OF THE FOCI OF AN ELLIPSE

The easiest way to think about finding the foci is to imagine an ellipse and to think about a few special cases. If we can prove a relationship for a few special cases, we can then reason about how to generalize that case to be sure it will always work. The distance from the center of the ellipse to each focus is c. The length of the semimajor axis is a and the semiminor, b. This situation is shown in Figure 11-26.

At point 1, the distance from the focus on the left is

$a + c$.

The distance from the focus on the right is

$a - c$.

The sum of the two is

$a + c + a - c,$

or just 2a.Now we know that the sum of the distances from each point on the ellipse to the two foci is 2a, if that has to be true everywhere.

But that still doesn't tell us the actual value of c. For that, we need to use point 2, on the y axis, since our special case ignores anything in the up and down (in our sketch) direction. At the point labeled "2" in Figure 11-26, the sum of the distances (dashed lines), by the Pythagorean Theorem, is .
$2\sqrt{c^2+b^2}$

We also know from what we just figured out with point 1 that this distance also has to be 2a. So we can set them equal to each other:

$2\sqrt{c^2+b^2} = 2a$

Square both sides and divide by 2 to get:

$c^2+b^2 = a^2$, or
$c^2 = a^2-b^2$

If you are used to looking at the Pythagorean Theorem that may bother you, but in math, the letters a, b, and c are used for many things. Unfortunately, in this case, it is in a way that fights our Pythagorean instincts from earlier chapters.

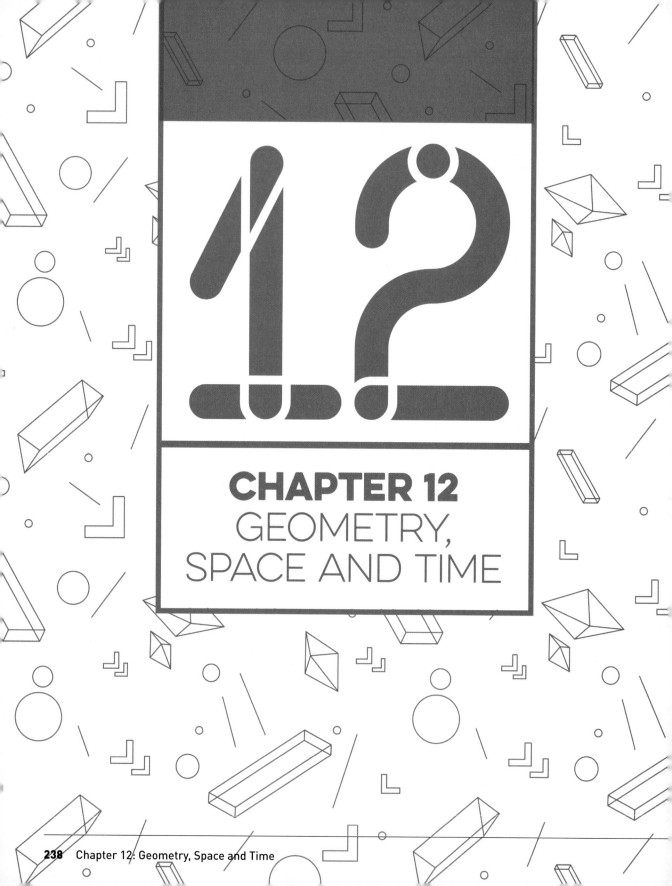

CHAPTER 12
GEOMETRY, SPACE AND TIME

In this chapter, we will look at astronomy and physics ideas that use the geometry concepts and their applications we've learned in other chapters. Specifically, we develop tools for you to be able to estimate the local time of day or time of year based on observing the sun, building on our work in Chapter 7 finding our latitude and longitude. Sundials of progressively increasing sophistication go back to prehistory, and we will show you some basics and point to ideas to go beyond.

If you can handle some delayed gratification, we also describe a means of tracking the sun's position for a year at a set time of day. We'll find that the sun (or, equivalently, its shadow) makes a sort of lopsided figure-8 over the course of a year called the *analemma*. Like everything else scientific in the late medieval period, early knowledge of it was shrouded in intrigue, but you can replicate it on a sunny patch of driveway. The analemma captures a lot of information about Earth's orbit around the sun over the course of a year, and we will see how to interpret our observations.

THE EQUATION OF TIME

In Chapter 7, we talked about the Equation of Time, which isn't really an "equation" the way we normally think of one. It is more like a correction that we have to apply if we want to know the time (or our position) based on where the sun is on a given day. One fallback for us modern folks, as we saw in Chapter 7, is to estimate from the graph in the Wikipedia article "Equation of Time" or do an internet search on "sundial correction." However, now that we know a little more geometry, can we be more intelligent consumers of these sources? And, at the same time, can we be more appreciative about how observant some of those people who lived a thousand or two years ago were?

First, let's think about Kepler's second law (Chapter 11), which says that, if we drew a line from a planet going around the sun, the line would sweep out an equal area of an orbit in equal time. One measure of how much an ellipse is different from a circle is its *eccentricity*, often called *e* (like many other things in math, unfortunately). For an ellipse with semimajor axis a and semiminor axis b, the eccentricity is:

$$e = \sqrt{1 - b^2/a^2}$$

It turns out that the earth's orbit has an eccentricity of 0.01671. So it is not a lot different from a circle, but different enough to be observable. If we square both sides and rearrange a little,

$$(0.01671)^2 = 1 - b^2/a^2 \text{ and thus } 0.99986 = b/a.$$

To put it another way, the shorter axis of our orbit is about 99.986% the size of the longer one.

Isaac Newton figured out an equation called the *vis-viva equation* that will let you figure out the speed at any point in an orbit. We won't wade into the algebra for that here, but it turns out that we get closest to the Sun in early January of each year (called perihelion) at which time we are clicking along our orbit at about 30.29km/s. We are farthest from the sun (aphelion) in early July, and going about 29.29km/s (according to Wikipedia, "Earth's Orbit."). In 2023, the dates are January 4 and July 6 respectively. The website **https://www.timeanddate.com/** is a very good source for data like this.

The implication is that we spend a little less time hanging around the part of the earth's orbit around the December solstice than we do around the June solstice. Note, though, that the way the tilt of the earth's axis happens to line up with the orbit is such that the time when we are closest to the sun does not line up with when we will see the shortest and longest days of the year. Also, we are very slightly closer to the sun while it is winter in the northern hemisphere.

The bottom line is that the effects of the tilt of the earth's axis, plus the differences in the earth's speed around the sun at different times of the year, sometimes add up and sometimes subtract from each other. They cancel out four times a year. The dates vary plus or minus a day or so, because the earth's orbit is also not a round number of days, and there are several effects (some of the biggest of which we will talk about in this chapter) that happen out of sync with each other. That said, the dates when the Equation of Time is zero are approximately April 15, June 13, September 1, and December 25. The rest of the time, the correction can be as much as 16 minutes one way or the other; in the next section, we see what the variation looks like.

There are some other factors too, like the effects of other planets tugging on the Earth-Moon-Sun system and a variety of other wobbles. To figure out the Equation of Time in detail is a big job. For the most part, only people who have to navigate spacecraft to where they are going and perform similar tasks will need to figure out the details.

As a side note, read the definition of the sign of the correction carefully if you are reading the Equation of Time from a table. The conventions are not consistent everywhere. Some people define a positive number as mean time minus solar time, and others reverse the order. We describe the convention where February is negative, where you would need to observe the sun at 12:16 on your clock to get a noon reading.

THE ANALEMMA

Other than intellectual curiosity, until recent times, there weren't a lot of reasons to need to know the time to very high resolution. That all changed when people started to navigate across oceans. Thus, it is not a coincidence that a lot of interest in better timekeeping and globe-spanning sea travel came along about the same time (in the 1500s). We went over the link between a good timepiece and being able to determine your position in Chapter 7.

There was also a desire to know when the spring and fall equinoxes were (the first day of spring and fall, respectively). People wanted to know when to plant crops, or when to celebrate holidays. Thus some monks, and scientists with patrons, started to do more thorough observations in the late 1500s and early 1600s. A new calendar, the *Gregorian calendar*, was introduced in 1582 as a result of some of these observations. Yet it was still an era with inherent contradictions for astronomy since it remained heretical in some countries to say the earth went around the sun.

As scientists gathered more and more evidence that the earth went around the sun, some observers in the late 1600s (like the Italian/French astronomer Jean-Dominique Cassini) began to take careful observations of *meridian lines*, which mapped the height of the sun above the horizon. Cassini worked with a pinhole in a cathedral at San Petronio in Bologna, Italy. This pinhole projected a disk of the sun on a measurement line on the cathedral floor. This gave him good information on how the sun moved over a year along a straight line, and other data that confirmed Kepler's ideas about how planets moved around the sun.

THE GEOMETRY OF THE ANALEMMA

As various instrumentation began to get better (including clocks that were not dependent on the sun) people began to observe that if you marked the position of a shadow at the same clock time each day, over the course of a year it would not mark a straight up-and-down line. It would instead create a lopsided figure-8. This is called the *analemma* (Figure 12-1).

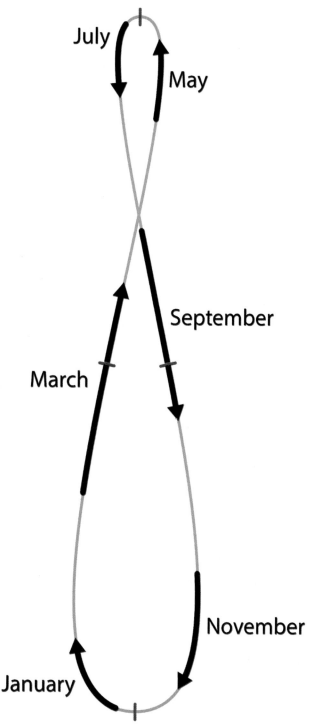

July

May

September

March

November

January

FIGURE 12-1: The analemma in the sky (North on top in the northern hemisphere).

The analemma is essentially a graph on the sky of the Equation of Time, drawn out by the position of the sun every day at one fixed clock time over the course of a year. Clock time (sometimes called "mean time") measures the long-term average over the year of the length of a day. On average, the sun will be at its highest point in the sky near noon, corrected for longitude (Chapter 7). However, sometimes the sun will be a bit further east or west than average at this time of day, and this creates the east-west part of the analemma.

The vertical part of the analemma is the seasonal elevation change, which we talked about in Chapter 7, when the sun appears lower in the sky in winter than it does in summer. However, even if the earth's orbit were perfectly circular, the analemma would be a symmetrical figure eight, not a straight line. The reason for this is a little subtle, and best seen in animations. It comes about because the earth's tilt (which an astronomer would call the earth's *obliquity*) makes the actual path of the sun in the sky take a little more or a little less time than the imaginary average path of the sun. For an excellent visualization look at the "obliquity" animation at **http://analemma.com/obliquity.html** or that site's associated phone app.

The difference between the top and bottom of the analemma mostly comes from the fact that the earth's orbit is an ellipse, and so at some parts of the orbit the earth will move a little farther along its path than it "should" on average (by Kepler's second law). The sun will seem to arrive a little later than it should (and thus be a bit east). The sun is behind where it should be on the left-hand side of Figure 12-1, and ahead on the right side.

The biggest excursions from the zero point are around Valentine's Day (-14 minutes) and a little after Halloween (+16 minutes). There are smaller excursions in mid-May (shy of +4 minutes) and mid-July (-6 minutes). On April 1, when the Equation of Time is about -3 minutes by this convention, you would measure the sun's position at 12:03 PM on your clock to measure solar noon. (Subtract the 3 minutes from clock time to get solar time.)

Since we want to observe this difference with our analemma drawing, we do our measurements at the same clock time every day (not adding or subtracting the Equation of Time). In fact, you must be as precise as possible. We watch our phones and make our mark just as 11:59 AM turns to 12:00 noon each day. The effect you are measuring is small, and even a minute or so one way or the other makes for some visible jiggling around on the curve.

The two parts of the figure 8 are not symmetrical because of the way these two effects interact, adding on the winter lobe and acting in opposite directions in the summer lobe. The analemma shown in Figure 12-1 is a bit stylized; the right and left aren't quite symmetrical either. The time of day at which the analemma is measured and the latitude will distort this figure somewhat as well.

MEASURING THE ANALEMMA

With some planning and forethought, you can measure an analemma yourself by measuring the shadow of a fixture that will stay put for at least a year, and that will throw a distinct shadow when illuminated by the sun from various angles. The area where the shadow lands will need to be accessible and clear of snow and ice most of the time, and you ideally want an area that is as flat and horizontal as possible.

You can measure the analemma at any time of day that will have sunlight at your latitude all year long, but away from noon the analemma will tilt significantly and might get quite large. We are in the process of marking one near Los Angeles, California at noon Pacific Standard Time.

We used the corner of the rain gutter on a south-facing wall of a garage as our shadow reference point (Figure 12-2). We started in mid-September last year, and the winter solstice is on the right-hand end. This photo was taken in late March. The curve is beginning to bend inward to cross over in mid-April (over on the left side of the photo).

We've realized it is going to go into the grass on the center-left of this photo

FIGURE 12-2: Analemma traced from the shadow of the corner of the rain gutter (upper left of photo).

a little, so we will put out some plywood for that week or so. Our September start is the open end of the curve farther from the garage door. South is to the left. The brick accent lines on the driveway are nearly due north-south and east-west.

We set a recurring phone alarm a few minutes ahead of time, and one of us goes out exactly at noon Pacific Standard Time (1 PM Pacific Daylight Time) as often as possible. (By the way, if you are not in the United States, "Daylight" Time is what you might call "Summer" Time.) We mark the shadow of the corner of the rain gutter (Figure 12-3) with washable chalk.

Whenever rain was predicted we put down a bit of white electrical tape just inside the track made so far, since the chalk would wash away. After it is done raining, we re-draw in the curve using the tape pieces as a guide. (You can see the tape as white strips in the analemma photos). Electrical tape stuck pretty well to the driveway if we brushed the concrete off first. Of course, we did not have snow to deal with, or, for that matter, very much rain in the winter and spring we were working on this.

Figure 12-4 shows the analemma from a different angle. The blue arrow

FIGURE 12-3: Marking the shadow at noon each day.

FIGURE 12-4: The analemma viewed from the south

on the concrete points to true north. (The smaller arm on the blue arrow is magnetic north, as discussed in Chapter 7.) It's not necessary to mark it every day, but you do need to be consistent in making your mark within a minute or less of your chosen time since the shadow moves quickly. Finally, note that the shadow of the analemma is inverted from how it appears in the sky. When you read about it, notice whether the writer is talking about the appearance in the sky or as a projection on the ground.

You could also take a time-lapse photo of a shadow on the ground, too, instead of marking it. If you have a security camera, perhaps you could save a noontime frame every day and then combine them with time-lapse software. You'll need to find a place that you can mark for a year, and consider your local weather. If you have a south-facing window (north-facing, if you are in the Southern Hemisphere) and the sun will shine on the sill even at the highest point in summer (see Chapter 7 to figure out those extremes) you might be able to do a miniature one on a windowsill. You will have to experiment with a dot on the window or some other small object to cast the shadow. A school playground using the shadow of the top of a flagpole might be a fun place to do it, with a mark made at the start of lunch or recess on school days. Or you can do a respectable analemma marking it once a week, so it could be a weekend ritual.

We have found it a very tangible way to see how the sun moves in the sky over the seasons. It helps in thinking about why the earliest and latest sunrise and sunset are not at the solstices. The time of sunrise is affected both by the length of the day and by the offset of the middle of the day. A day in May can be the same length as one in August, but sunrise and sunset are both earlier in May.

There are many readily available daily tabulations of the Equation of Time available if you prefer to just sketch one based on that data. Notwithstanding that, we encourage you to play with marking the shadow of the analemma for a time to see how the passage of time plays out in the sky (and shadows).

Incidentally, think about the analemma at the South pole. What would someone observe? Search online for "analemma south pole" for interesting time lapses of the answer.

Finally, if you want to read the Equation of Time from the results of your year-long project, you will need to show a line for the times when the correction is zero, and draw a scale that has the appropriate maxima at the appropriate places.

ANALEMMAS ON OTHER PLANETS

If you were to mark the analemma on other planets, you would come up with different figures depending on the axial tilt of the planet, how elliptical its orbit was, and any other secondary factors. Mars, for instance, has an axial tilt a little bigger than that of the Earth (about 25°, versus the earth's 23.5°). However, its orbit is a lot more elliptical than Earth's is (the semi-minor axis is 99.6% of the semi-major one). This works out to a teardrop-shaped analemma, which has now been assembled from photographs by the Mars Opportunity rover. If you search online for "Mars analemma" you can check out various visualizations and analyses.

PATHS IN THE SKY OF OTHER OBJECTS

The word "analemma" usually refers to the path of the sun in the sky (of earth or another planet). People have also done time-lapse photos of the position of the moon in the sky, which they refer to as an "analemma of the moon." Usually, they allow for the rotation of the moon around the earth and do the photo later each day; even so, because its orbit is not a circle either, the path is complicated.

Other planets make more convoluted paths in the sky, because they, too, are moving around the sun in their own elliptical orbits. If you were to photograph a planet, say, Mars, every night, you'd almost certainly see it make a loop in the sky, as the Earth overtook the red planet in its orbit. The way the planets wander around the sky was some of the evidence that earth did, after all, revolve around the sun. The word "planet" even comes from the Greek word for wanderer.

MAKING A SUNDIAL

Sundials have been around for a very long time, and for much of history it didn't matter if the time they displayed was 15 minutes off, since people didn't know about, or bother with, the correction for the equation of time. In this section, we are going to talk through making an extremely simple sundial. We will use paper and a 3D-printed piece, but you could use just cardboard and paper, too. We will talk through the Northern Hemisphere build here. If you are in the Southern Hemisphere, interchange "north" and "south" in our directions.

CREATING THE GNOMON

To start our observations, we first make a somewhat more sophisticated gnomon than our stick in Chapter 7. A good way to avoid distortions as we track the sun is to create a gnomon that is parallel to the earth's axis. The OpenSCAD model **sun_dial_gnomon.scad** creates a triangular object, the top of which is parallel to the earth's axis if it is pointed at true (not magnetic) north. There are several variables you will need to define. In particular, you need to know your latitude, in degrees. Change the value of latitude to equal your latitude, to at least one decimal place.

```
length = 100; //length of the base of the triangle, mm
base = 30; // width of the base of the triangle, mm
latitude = 34.1; //latitude of user, degrees
thickness = 2.2; //thickness of the piece, mm
```

FIGURE 12-5: Gnomon (latitude angle, and due south, on the right)

If you don't have a 3D printer, create a right triangle from cardboard or foam core with an angle at the base equal to your latitude, and create a base it can stand on. In other words, make an equivalent in cardboard of the gnomon created by **sun_dial_gnomon.scad.** Figure 12-5 shows the gnomon, created for 34.1 degrees latitude.

We marked our example in early April, where the Equation of Time is about -3 minutes. That means that the sun gets where it needs to go 3 minutes late. So, to make up for that, we would observe 11 AM at 11:03, noon at 12:03 PM, and so on. If you did this in, say, mid-October when the Equation of Time is about +12 minutes, you would make your observation 12 minutes before the hour (11:48 AM for clock time of noon). We did this in Daylight Saving Time, so the sun is near its maximum at about 1 PM. We are however at 118 degrees longitude, so (as discussed in Chapter 7) the solar time will be about 8 minutes earlier than the clock time hour. So we would expect our highest sun at about 1:03 PM - 8 minutes, or 12:55 PM Pacific Daylight Time.

Once you have made your gnomon, take four pieces of 8.5x11 paper (or equivalent) and tape them together down their long ends. Don't overlap them. Lay them side by side and then put down tape, so they don't get crooked. (In other words, end up with one 34x11 taped-together piece of paper.) There's nothing magic about the size. You just want something big enough that the longer shadows will fit, which might be quite large in some latitudes. If it is excessive, make the gnomon smaller (preserving the angles).

FIGURE 12-6: Gnomon on paper, marking 8 AM. True north is marked with an arrow.

Put your gnomon on the taped seam in the middle of the four pieces, and mark where it is so you can replicate the setup easily. The gnomon's angle which is the same as your latitude should point at the sun (south). Use your compass to line up the seam between sheets of paper, or at the edge, so that it is aligned with a north-south line. (Figure 12-6).

Then, at your local clock time modified by the Equation of Time offset for the day, make a mark of where the shadow is at 8 AM. We strongly suggest using a pencil, not a pen, since you might discover something was a little off and have to re-do some points.

There's no need to adjust for where your longitude puts you in the time zone. If you are going to read the time in the same place you are doing this measurement, you are automatically calibrating it for your latitude and longitude. (You can see why, in the era of sundials, every town had its own private time zone.)

A day in May can be the same length as one in August, but sunrise and sunset are earlier.

Repeat this process at least every two hours to make your marks (perhaps with a few extra points near noon to get that area, where the lines are close together, as accurate as possible). If there is no longer direct sun where you started you can move the whole assembly, as long as you line it up with true north again and are sure the gnomon is in the same place on the paper

FIGURE 12-7: Marking time at Pacific Daylight Time 1:03 PM

(Figure 12-7). Be sure that the ground is as flat and level as possible where you do both observations. If you are going to make a sundial in something more permanent than paper, you might want to make several observations at each hour to be sure something wasn't off.

Once you have taken your observations, mark where the point of the sundial was (the point underneath where your latitude angle starts). Using a long ruler, carefully draw a line from this point through each of the observed points. Figure 12-8 shows one of these lines in red.

It might take you a few days to measure at enough times. Just be sure to allow for the changing Equation of Time offset of when you take your measurement. And don't be afraid to erase a point or two if you discover you had some stuff set up wrong, or if you did the Equation of Time backward. You'll see some erased points on our final graph if you look closely!

Once you've drawn your lines, you will have something that looks like Figure 12-9. (Draw up to and through your data point, to the edge of your paper.) You can see that the sun was so bright that it faded the paper around the edges of the gnomon after just a few days of standing outside in the California spring sun. This makes it easy for us to see where the gnomon was, though.

After you do that for all the data points, you now have a minimalist sundial (Figure 12-10). You could use this as a template to make something more permanent, perhaps by measuring the angles and re-drawing them on

FIGURE 12-8: A line running from the origin of the gnomon through the 6 PM marking.

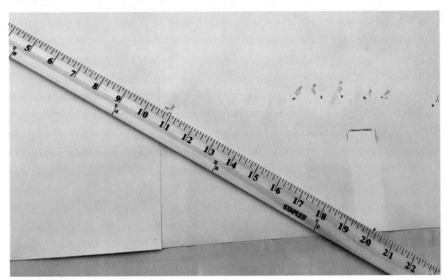

FIGURE 12-9: Drawing the lines

something that will stand up to the elements better than a piece of paper.

USING YOUR SUNDIAL

To use your sundial, allow for the Equation of Time, and read the time based on where the shadow has fallen. If the Equation of Time is negative, take your reading the appropriate amount after the hour (e.g. 12:03 PM to look at the noon marking if the Equation of Time is -3 minutes.)

In winter, the tips of the shadow will cluster farther out from the center point,

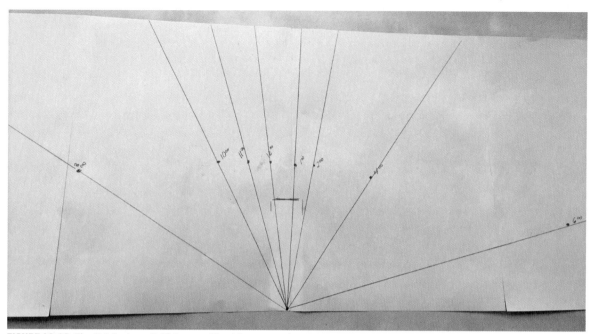

FIGURE 12-10: The finished sundial (shown without its gnomon).

since the sun is lower and the shadows are therefore longer. In summer, they will be closer in, but 3 PM should always fall pretty close to the 3 PM line (allowing for Daylight Saving and the Equation of Time).

You can think of these radial lines in Figure 12-9 as the centerline of the analemma at that time of day at your longitude. The sun may be higher or lower along that line at that time of day, but it will still be nearly on that line at all times of the year. The angle of the gnomon lines up with the earth's axis of rotation, and the sun always moves across the sky at 15° per hour (360°/24 hours). That models the north/south part of the analemma simply for us.

There's no simple way to model the east-west part of the analemma which is, as we have mentioned, a combination of several factors, so you're sort of stuck with a correction for the Equation of Time. If you cruise the internet looking at sundial designs you will come across a lot of elaborate attempts, mostly with varying lines for different times of the year. As a side note, the top of the shadow of the gnomon traces out a different hyperbola each day (a straight line on the equinox). There's a complicated construction for that which would take a while to get into, but at bottom, there is a focus and circular directrix involved, after a lot of secondary construction to get to them.

CLOCKWISE

You will notice that the sun's shadow moves around in a clockwise direction, and that noon is (more or less) right above the gnomon. You can see where the basic design for clocks came from originally!

OTHER SUNDIAL DESIGNS

If you want to, you can take the next step, consider how to make a more elegant or permanent version of this using wood or other materials. The website **https://equation-of-time.info/** has an amazing amount of data about exactly how to correct (and construct) sundials. Some people also like to post little tables of the Equation of Time on their sundials, so you can know whether to add or subtract minutes from the displayed time and how much.

If you do an internet search on making a sundial, you will see a huge number of options. (Caution: as with anything on the internet, some may be wrong!) The next step up from the very simple one we described here would be to create an *equatorial sundial*. Instead of a flat sheet of paper for the location of the tip of the shadow, these have a curved half-circular band tilted at the user's latitude. This can be more compact than the flat sundial and changes less over the course of a year (although it still needs the Equation of Time correction).

SUMMARY AND LEARNING MORE

In this chapter, we learned about how to use the sun's position (technically, the shadow it causes) to learn more about Earth's orbit and how it affects an attempt to tell time based on the position of the sun. We also learned about the practicalities of measuring the analemma over the course of a year. Finally, we explored creating a very simple sundial, and places to explore more. We were very impressed by the site analemma.com if you want to learn more.

In the next and final chapter, we will take a tour through an imaginary museum of geometry, and learn about the development of some architectural features as well as an assortment of projects that we hope you will find intriguing.

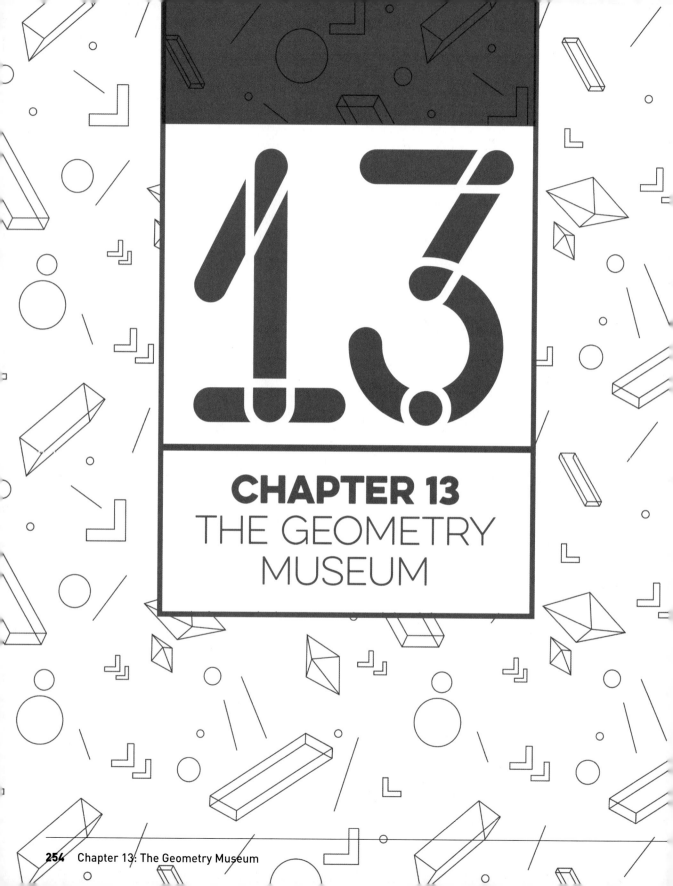

13

CHAPTER 13
THE GEOMETRY MUSEUM

We will wind up the book with projects that explore a few corners of the geometrical universe that we just think are cool. We call it the Geometry Museum, since we will walk you through a mix of history, architecture, and some pieces that are interesting for their own sake.

Architects have to know how to make buildings both strong and beautiful. Although today's architects have sophisticated CAD tools, medieval architects created spectacular buildings with bigger equivalents of the basic compass we used in Chapter 4. In this chapter, we will look at how some ideas we learned in earlier chapters have been applied historically to architectural problems, and how you can launch some projects of your own.

FIGURE 13-1: Stone arch at the ruins of Castell Dinas Brân in Wales. (Photo by Stephen Unwin)

First, we will explore an OpenSCAD model that allows you to build arches that are circular, as well as ones that are based on Reuleaux triangles (Chapter 4). Then we'll add arched windows and a door to the castle we did way back in Chapter 3. Finally, we introduce an OpenSCAD model of a Gothic arch window that you can add to and modify.

In Chapter 5, we started to explore why triangular structures appear so often in large-scale construction. In this chapter we learn that they are

FIGURE 13-2: Ruins of Valle Crucis Abbey, Wales. (Photo by Stephen Unwin)

called *trusses* and will suggest projects you can do to explore structures, using straws and elastic cord.

Finally, (as with all museum visits) we will end in the "gift shop". We will show you a few ideas that we think of as geometry puzzles—relationships that are just simply intriguing for their own sake. We hope that a little walk through our imaginary geometry museum will give you a lot of ideas for projects of your own that you can take to the next level.

ARCHES

Let's start our tour through our geometry museum by thinking about buildings that have stood for hundreds or thousands of years. Classical Roman buildings like the Coliseum (constructed over 2000 years ago) or medieval cathedrals like Notre Dame (which took about 100 years to build, finishing about 760 years ago) feature graceful *arches*, openings with curved tops, in their structure. Figure 13-1 shows an ancient ruined arch near Llangollen in Denbighshire, Wales. It is part of what remains of a fortress

3D Printable Models Used in this Chapter

See Chapter 2 for directions on where and how to download these models.

arches.scad
Creates two kinds of freestanding arches

tracery.scad
Creates a shape like a medieval church window

castle_advanced.scad
Adds arched windows and a door to the castle from Chapter 3

puzzlebox.scad and puzzletetrahedron.scad
These make a cube and its inscribed tetrahedron (scaled so the tetrahedron fits in the cube). More generally, puzzlebox.scad makes a hollow box missing one side which we will use for a couple of purposes.

edge_platonic_solids.scad
Prints one or more Platonic solids, scaled by the user input length of one edge (introduced in Chapter 3).

golden_ratio.scad
Creates a series of boxes with sides in a ratio called the golden ratio.

meissner.scad
Creates Meissner objects of constant width

revolved_reuleuax.scad
Creates surfaces of revolution from Reuleaux polyhedrons

Other supplies for this chapter
- Removable glue putty (e.g. UHU Tac, Blue Tack)
- 18 plastic drinking straws
- Ruler
- Elastic cord. We used 0.8 mm elastic cord, also called beading thread or crafting cord. You'll need about 5m total.

called Castell Dinas Brân (also known as Crow Castle, thought to be built in the 1260s). Figure 13-2 is what's left of an ancient abbey near the fortress dating to around the year 1201, called Valle Crucis Abbey.

How can a massive stone structure last for 1800 years while supporting the huge weight bearing down on these openings? We are going to explore arches mechanically and then artistically in this section, ending with a little renovation of our Chapter 3 castle to add some arched windows and a door. After that, we will learn about the stonework inside windows like those in Figure 13-2, called *tracery,* and try creating some of our own.

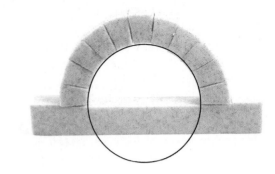

FIGURE 13-3: A circular arch

CIRCULAR AND GOTHIC ARCHES

In medieval times, building site instrumentation was limited. They did, however, have various ways of drawing circles. A circular arch can always be made by pinning one end of a rope and swinging it around that pivot. Shapes that can be made with just a rope or large drawing compass are practical to lay out, and if they are also strong, so much the better.

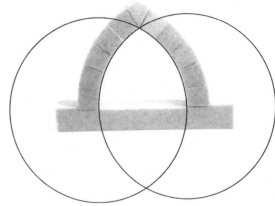

FIGURE 13-4: A Gothic arch

To help us think about a simplified version of stone arches, we've created the model **arches.scad**. It allows you to make a 3D printed model of two types of arch: a circular arch, and a Gothic arch. A circular arch is what it sounds like: an opening in a structure that is half a circle. To understand a Gothic arch, though, we need to go back to Chapter 4 and look at the construction of a Reuleaux triangle.

To make a circular arch, one just needs to draw a half-circle (Figure 13-3). To make a Gothic arch, one draws two circles of the same radius as each other. Each circle is centered at the base of the other arc (Figure 13-4). If we go back to Chapter 4, we'll see that to draw our Gothic arch we used part of the construction of an equilateral triangle. The arch is a Reuleaux triangle with one side flattened.

If you want examples in real buildings, Figure 13-1 shows a circular arch, and Figure 13-2, several Gothic ones. If you walk down any street in a modern city, you are likely to see many examples of each.

If a Gothic and circular arch have the same size opening at their respective bases, can you figure out the difference in their heights? (Ignore the thickness of the stones.) What about the perimeter of each arch, again ignoring the thickness of the stones? The answers are at the end of the chapter.

MAKING THE MODEL

The model **arches.scad** allows you to create arches (and a base for them) with various parameters. It makes individual "stones" appropriate to the arch. For the circular arch, all the bricks are the same. For the Gothic arch, all the stones are the same except for the one at the very top. That is called the *keystone*, and the geometry of the arch requires that it be different from the rest. Variables you can change, and their defaults, are:

- `size = 80;`
    - diameter of the arch opening at the base, in mm
- `w = 10;`
    - width of the bricks that make the arch, in mm
- `thick = w;`
    - height of the print, set to 0 for 2D printing on paper
- `segments = 11;`
    - Number of stones; use 10.01 for a small keystone (stone on top of a Gothic arch)
- `gap = 0.1;`
    - The gap between stones (to leave space for mortar, for example), in mm
- `separate = 0;`
    - The extra spacing between the stones for printing (to ensure that they don't stick together).
- `gothic = false;`
    - Make this **true** to print a Gothic arch, **false** for a circular one

The values we show as defaults worked well for us, but feel free to explore and play with them. The overall longest dimension of the model will be `size + 4 * w,` since the base is wide enough for the arch plus one stone's worth of extra base on either side. You will notice the base has a notch in it so that it can exert a little side force. To assemble your 3D printed arch, lay out the pieces flat on a table. Set aside the keystone if it is a Gothic arch.

Start at the bottom (on the base). Put a tiny dab of putty as mortar between the pieces as you go (Figure 13-5). We used UHU Tac removable putty, sometimes called "museum tack" and sold to hold down vases in an earthquake. Use something that will stay soft since you are not making a permanent attachment. In the next section, we will want to take it apart and put it back together.

For our build, we used a "marble" PLA so that it would even look like stone. This material has little flecks of darker-colored material mixed in to give it a stone-like appearance. This specific type of additive is generally not abrasive, but other filament additives may be. If you buy marble PLA filament, check to see if the manufacturer warns that it will abrade your nozzle. If so, be sure your printer has a hardened nozzle that can handle it.

FIGURE 13-5: Putting a bit of putty between the pieces

TESTING YOUR ARCHES

Print and assemble (with some putty) two arches with all the variables the same, except that one should be a Gothic arch (**gothic = true**), and one circular (**gothic = false**). Now, set your arches up vertically (Figure 13-6).

FIGURE 13-6: The two arches

Press your hand down on the top of each arch. What happens? (Figure 13-7) Most likely the circular arch collapsed with far less force than the Gothic one. The circular arch is an older invention, and the Gothic is an improvement. More force is transmitted straight down in a Gothic arch compared to a circular one.

FIGURE 13-7: How arches collapse (or not)

FIGURE 13-8: Closeup of one stone from circular and from Gothic arch models

A circular arch requires more support from the sides, and if you do this experiment several times, you will probably see stones get squeezed out to one side or the other as the arch fails. The design of a Gothic arch keystone is particularly clever. As the weight of the building above pushes down on the keystone, the stones on either side want to push upward and out, counteracting some of the downward forces. Of course, those stones are supporting a lot of weight themselves, which gets extremely complex very fast.

As a side note, "Gothic" was a name applied by architects at the time who felt the style was barbaric, so they named it after the marauding Goths. If only they could see which buildings have lasted for a thousand years!

Light and slippery plastic pieces with elastic putty for mortar aren't a perfect simulation of the physics in the real cathedrals. There, the mortar would be rigid as well. You can definitely see why that long-lost person who perhaps did an equilateral triangle construction in their geometry class probably got pretty excited when they created a small prototype, though.

One thing to notice is the shape of the pieces (Figure 13-8). They are wider on the outside than on the inside of the arch. That means that a force from above will want to wedge the pieces in tighter. For the circular arch in particular, if a piece had parallel sides it would want to just fall out.

Arches do experience some sideways outward force. That is why the base of this model has a small ridge to hold the arch in place. More elaborate structures were used in medieval times. Sometimes these were enormous stone piers, but sometimes these were also arches, called *flying buttresses*, which were engineering marvels themselves.

Also, we have talked about how 3D printers can only create overhangs of 45 degrees or so. Steeper than that, and the plastic may sag down. If you look at the overhang angles of the Gothic and circular arches, you can see that the overhangs on the Gothic arch never get as steep as those on the top of a circular arch.

PAPER (OR CUT-OUT) ALTERNATIVES

If you set **thick = 0,** you can export a .dxf or .svg 2D file from OpenSCAD instead of an STL. As we described in Chapter 9 when we did this for nets, you can then print this on paper. You won't be able to do the tests we described with paper, but you could use it as a pattern overlaid on other materials. For example, you could glue the paper onto heavy foamcore, cut out the pieces, and play around with that. Or you can adapt the other parameters a little so that they will work on your laser cutter (for instance, you might want to use a nonzero value of **separate** to spread the pieces out a little). This pattern is not intended for anything structural, but you can have fun with it for scale models!

RENOVATING THE CASTLE

At the end of Chapter 3, we drew a windowless and entry-less castle, which obviously is of limited utility unless you are a dragon that can fly in and out. Now, we will add those features in the form of windows with a rectangular bottom and arched top. This version is in the file **castle_advanced.scad.** The doors and windows are going to be created with a *module* called **arch**.

Modules are pieces of code that you want to use over and over. You can use different values of parameters each time you use the modules. We say that we are *passing values to the module*. This module takes in a three-dimensional set of numbers (a vector—see Chapter 2) and a parameter to tell it whether we are making a Gothic or circular arch.

Let's walk through how it works. These values are passed to the module:
- **size** ([x, y, z])
    - the x, y, and z dimensions of the arch, where z is the thickness of the arch (it is generated in 2D in the x-y plane) and x and y are the dimensions of the rectangle below the arched part. The arch is generated with a base diameter of the x value. **size[0]** is the x value, **size[1]** is the y value, and **size[2]** is z.
- **gothic = true** (or false).
    - Whether it is a Gothic (**true**) or circular (**false**) arch-topped window.

The **arch** module is tricky. It exploits the fact that OpenSCAD models in 2D are often faster to create than 3D models. First, the code checks the z value (**size[2]**) to see if you are trying to produce a 2D or 3D shape. If you provide a non-zero value, it uses **linear_extrude()** to extrude the arch shape to the specified thickness. When you do this, the module gets the arch shape by calling itself, a process known as *recursion* (discussed in Chapter 2). It passes along all of the values that were passed to it, except that this time it sets the thickness to zero.

When the z value is zero (or if you only provided x and y values, and there is no **size[2]**), the code follows the other branch, which generates the 2D arch shape. This code will generate either a Gothic or circular arch, with a base width equal to the x value (**size[0]**), with a rectangular section below it equal to the y value (**size[1]**). The rectangular section overlaps the circle used to make the circular arch, or the two intersected circles for the Gothic one, and an **intersection()** is used to trim off anything that extends below the rectangle. Here is the module:

```
module arch(size, gothic = true) {
  if(size[2]) {
    linear_extrude(size[2]) arch([size[0], size[1], 0], gothic);
  } else intersection() {
    union() {
      if(gothic) intersection_for(i = [-1, 1])
        translate([i * size[0] / 2 , 0, 0]) circle(size[0]);
      else circle(size[0] / 2);
      if(height > 0) translate([-size[0] / 2, -size[1], 0])
        square([size[0], size[1]]);
    }
    translate([-size[0] / 2, -size[1], 0]) square([size[0], size[0] + size[1]]);
  }
}
```

To add doors and windows to the Chapter 3 castle, we will need to add a row of windows to the side and back walls, and one door and flanking windows to the front. We have to rotate and translate these appropriately, too, so they appear in the various walls of the castle. (Remember that they started out in the x-y plane, but the castle walls are in the y-z or x-z planes.)

Finally, we will need to subtract these from the original castle using **difference()** to make it negative space. Here is the entire model, including those additions (available as **castle_advanced.scad**). When all is said and done, you will have the castle shown in Figure 13-9.

```
wall = 100;
height = 40;
thick = 4;

$fs = .2;
$fa = 2;

difference() {
  for(side = [0:1:3]) {
    rotate(90 * side) {
      translate([-wall / 2, wall / 2, 0]) {
        translate([0, -thick / 2, 0]) {
          cube([wall, thick, height]);
          for(crenellation = [0:thick * 2:wall - thick]) {
            translate([crenellation, 0, 0]) {
              cube([thick, thick, height + thick]);
            } //end translate
          } //end for
        } //end translate

        cylinder(h = height + thick, r = 1.5 * thick);
        translate([0, 0, height + thick]) {
          cylinder(h = 3 * thick, r1 = 1.5 * thick, r2 = 0);
        } //end translate
      } //end translate
    } //end rotate
  } //end for

  // All of what follows will be subtracted from the castle
  // Now create the rows of windows on sides and back
  for (a = [1:1:3]) rotate([90, 0, a * 90])
    for(i = [-2:1:2]) translate([i * wall / 6, 20, 0]) arch([5, 10, wall], true);
  // Next create the front door and a window on each side of the door
  rotate([90, 0, 0]) {
    for(i = [-2, 2]) translate([i * wall / 6, 20, 0]) arch([5, 10, wall], true);
    translate([0, 20, 0]) arch([25, 25, wall], false);
  }
}
```

FIGURE 13-9: The castle with windows and a door

FIGURE 13-10: Tracery

```
// Module that creates the arches
module arch(size, gothic = true) {
  if(size[2]) {
    linear_extrude(size[2]) arch([size[0], size[1], 0], gothic);
  } else intersection() {
    union() {
      if(gothic) intersection_for(i = [-1, 1])
        translate([i * size[0] / 2 , 0, 0]) circle(size[0]);
      else circle(size[0] / 2);
      if(height > 0) translate([-size[0] / 2, -size[1], 0])
        square([size[0], size[1]]);
    }
    translate([-size[0] / 2, -size[1], 0]) square([size[0], size[0] + size[1]]);
  }
}
```

GOTHIC TRACERY

Glass for windows was very expensive and not very strong until just a few hundred years ago. Before that, large windows were broken into panes by structures called *tracery*. Then, the open spaces would be filled in with stained glass. The model **tracery.scad** allows you to create designs in the spirit of medieval cathedral windows. Its outer boundary is a Gothic arch, and the shape is repeated to segment the interior with other, smaller arches (Figure 13-10). In medieval times, the long skinny openings would have probably needed some cross-bracing, but we will take a little artistic license here.

MAKING THE MODEL

Here are the parameters you can set in the **tracery.scad** model.

- `size = 60;`
  - Overall height of the piece, mm
- `wall = 1;`
  - The thickness of the walls, mm
- `depth = 5;`
  - Depth of the tracery walls, mm
- `backing = .5;`
  - The thickness of a backing plate, mm.

If you set **backing = 0** the tracery will be open, as in Figure 13-10. If you want to print the tracery in a transparent material or perhaps paint it, you would set **backing** to a nonzero thickness. The rest of the model is a bit like a Swiss watch; it will be tricky to change too much of it without distorting it. Figure 13- 10 showed it without backing; Figure 13-11 shows it with it (and the one without backing behind it, so you can compare and see how translucent you can make a thin backing).

FIGURE 13-11: Tracery with backing

THE TREFOIL

The three-lobed feature in the upper part of the arch in Figures 13-10 and 13-11 is called a *trefoil*. It appears often in medieval churches, although more commonly with the two circles on the bottom rather than on the top, as we have fit it in here.

To generate a trefoil yourself (Figure 13-12):

- Sketch a circle (blue).
- Draw three circles of the same radius as the original one, 120 degrees apart, with their centers on the first circle (red circles).
- If you remove the interior intersections you will get a trefoil (marked in black in Figure 13-12).

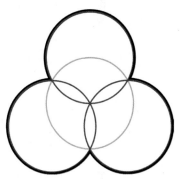

FIGURE 13-12: Circles comprising a trefoil

PRINTING CAVEATS

Printing without the backing (backing = 0) can be a challenging print, depending on your printer and slicer settings. If you like to make your first layer width a lot larger than subsequent layers (as we recommend in general), you may have an incomplete first layer and problems afterward. Simulate the first layer in your slicer, and if you see just a few spots instead of a full tracery back off on your first layer width. We used 150% of our regular line width (as opposed to our usual 200%).

FIGURE 13-13: The Mathematical Bridge, Cambridge, England (Photo by Stephen Unwin)

Alternatively, use 0.5mm for backing and your usual line first line width, and your print will look like Figure 13-11 versus Figure 13-10. In either case, the print is pretty delicate. Pry it off the platform gingerly and don't tug on those delicate, thin parts.

TRUSSES

In Chapter 5, we looked at how a triangle is stronger than a square. We also saw how adding a cross-bar to a square makes the square a lot harder to squish. Thus, it may not be surprising that we see triangles everywhere in bridges, buildings, and other structures. A structure made of triangles is often called a *truss*. In this chapter, we are going to try a 3D version of the 2D exercise we did in Chapter 5.

The key thing about a truss is that the joints are free to rotate, but the elements (the straws) are stiff. It turns out that this means that every element is only pulled on (*tension*) or experiences forces pushing it in at both ends (*compression*). The elements are not being bent, though. A good example is the Mathematical Bridge at Queens College of Cambridge University in Cambridge, England (Figure 13-13).

This bridge was designed in 1748 by William Etheridge and built in 1749 by James Essex the Younger. According to the college's website (**https://www.queens.cam.ac.uk/visiting-the-college/history/college-facts/mathematical-bridge**), the design uses many spans of lumber that are much shorter than the total span. This truss design

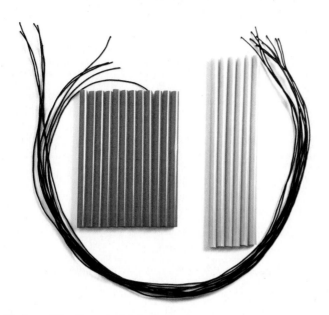

FIGURE 13-14: Assembling the materials for trusses

(called a *voussoir arch*) is constructed so that gravity causes all the elements of it to be in pure compression. Supposedly any element could be removed without affecting the rest of the bridge, although the college notes dryly that this has never been tested.

Trusses are relatively easy to analyze (although beyond what we can get into here) and you can look up how to calculate the forces on an idealized truss in engineering textbooks. But let's try making a 3D truss and see how it improves upon a structure that is made up of lots of square elements.

PREPARATION

First, take some stretchy cord (the same as in Chapter 5) and pre-cut yourself ten strands about 500mm long. These pieces will be going around the perimeters of a square 100mm on a side and a triangle 140mm on a side, with enough left over to tie a knot.

Next, gather 18 drinking straws. Plastic straws are the easiest to cut cleanly, and being pretty precise is important. The straws need to have a big enough diameter that two threads of your stretchy cord will fit through easily. If you are using bendy straws, be sure to cut off the bendy part. Cut them as follows:

- Twelve 100mm (about 4 inches) long pieces
- Six 140mm (about 5.5 inches) long pieces

You must be careful that all the straws in a group are identical, and that the longer straws (blue in our example) are 1.4 times the length of the shorter ones (red in the photos that follow). Now you should be ready to start (Figure 13-14).

TETRAHEDRON

First, let's try making a tetrahedron. A tetrahedron has four identical triangular sides. We will use the six longer (140mm, blue) straws we pre-cut, and four of the pieces of cord. Take three of these straws and run a piece of cord through them (Figure 13-15) to make a triangle.

Tie off the triangle (Figure 13-16), but don't cut off the excess cord just yet.

Now, take another piece of cord. Run it through one new straw, then one side of the triangle, then up through another piece of straw (Figure 13-17).

Tie off these two new pieces to create a second triangle (Figure 13-18).

Take one last piece of cord. Run it through the final 140mm straw, then through one of the outer straws of each of the two triangles you just created (Figure 13-19).

Swing the last piece up and tie off its loose end to make a tetrahedron (Figure 13-20).

At this point, the straws form a tetrahedron, but you still have one triangular face without a loop of cord. Thread one more piece of cord through these three straws (the ones with only one piece going through them) to complete the build. If you don't add the last cord, the tension on the joints will be uneven, and it may not work well when we merge it with the cube later on. Tie all ends securely and clip off the excess. If you cut too close to the knots, they may come undone, so leave at least a centimeter or two on each one. (Figure 13-21).

Finally, you can move the knots so that they are inside the straws, which will look nicer (Figure 13-22). Each joint should look like Figure 13-22, with two cords running through each straw.

Because all of the faces are triangles, this structure will be pretty rigid, even though the joints are just made of string. How do you think a cube would behave? Let's try that next.

FIGURE 13-15: Assembling a triangle

FIGURE 13-16:. Tying off the triangle

FIGURE 13-17: Adding the next two sides

FIGURE 13-18: Making the second triangle

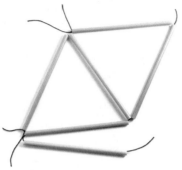

FIGURE 13-19: Adding the last piece

FIGURE 13-20: The tetrahedron

FIGURE 13-21: Trimming off the ends

FIGURE 13-22: Pushing the knots into place.

CUBE

Next, we are going to make a cube, and demonstrate how it and the tetrahedron we just made can be combined into a truss (and why you want to do that). To get started, collect the twelve shorter (red) straws and six pieces of cord. First, take four of the straws and run a cord through them to make a square. Tie off the square. (Figure 13-23).

Now take a second piece of cord, and run it through three more straws, plus one side of the square we just made. Tie it off to make a second square (Figure 13-24).

Now we will add two more sides. Run cord through the straws forming one long side of the two-square polyhedron. Add another straw on each end. (Figure 13-25)

Next, tie together the two pieces you just added. The outside of the shape will become a hexagon, and there will be three straws going to the center, creating something that looks like a projection drawing of a cube (Figure 13-26).

Next, run a cord into one of the corners of the hexagon that doesn't have a straw connecting it to the center. Thread it through that straw and the next perimeter straw, so that it will come out of another corner that is not connected to the center. Add two more pieces of straw (Figure 13-27).

Now, tie together the two pieces you just added. At this point, it may start to pull itself into a 3D shape, and you'll be able to hold it up to make a cube that is missing one edge. (Figure 3-28).

Your shape now has three straws meeting at each of its corners, except two, and you'll use your last 100mm straw to connect those two. Your last two pieces of string will both go through this straw, with each one making a loop around one of the two remaining faces. As with the tetrahedron, when you're done, every straw will have exactly two pieces of cord going through it. (Figure 13-29). This once again makes certain that the joints are all equal.

Finally, tie off the loose strings. You will now have a cube (Figure 13-30). However, you will discover that the cube doesn't want to be a cube. It will just squish from side to side with no real strength whatsoever. Cut off the excess cord and slide the knots inside the straws as before.

FIGURE 13-23: Making a first square side.

FIGURE 13-24: Adding the second square

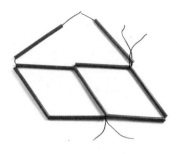

FIGURE 13-25: Adding two more sides

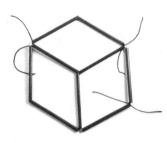

FIGURE 13-26: Tying together two pieces we just added.

FIGURE 13-27: Adding two more edges to the future cube.

FIGURE 13-28: Tying together these two pieces

FIGURE 13-29: Adding the last piece

FIGURE 13-30: Our final cube

FIGURE 13-31: Putting the cube in the tetrahedron

FIGURE 13-32: Cube vertex joined with tetrahedron vertex

FIGURE 13-33: Finished cube

COMBINE CUBE AND TETRAHEDRON

The cube really will not even stand up on its own. What happens, though, if we were to combine the tetrahedron and the cube? If we think geometrically, we see that the tetrahedron has six edges. The cube has six sides. What happens if we try to line up the tetrahedron so that one edge is a diagonal of each face of the cube?

First, take the cube and squish it into a four-pointed figure, like the inside of a tetrahedron. Push it into the tetrahedron as shown in Figure 13-31 so that one corner of the collapsed cube sticks out through each of the four sides of the tetrahedron.

Now pull out the vertices of the cube so that cube vertices that are squished into the center each come out and press out on a vertex of the tetrahedron (Figure 13-32). Stretch the cord a bit as you do so, so that the cube's corner goes through the tetrahedron's corner, to the outside of the shape. This will link the two corners together.

Finally, you will see that the tetrahedron sides are now diagonal supports of each side of the cube, just as we saw in Chapter 5 when we added a diagonal support to a square (Figure 13-33).

We made the tetrahedron sides 1.4 times longer than the sides of the cube. If you look at any triangle created by two cube edges and one tetrahedron edge, they form a right triangle with two sides the same (let's call that equal to 1) and therefore the Pythagorean theorem says that the diagonal, the hypotenuse of the triangle, (edge of the tetrahedron) will be the square root of $1^2 + 1^2$, or , $\sqrt{2}$ which is roughly 1.41.

We can see that the cube is now very stable and very strong. Variations on this theme are called a box truss. Trusses can be made of nothing but triangles. Here is an example of a complicated truss that is made the same general way we just made our truss and tetrahedron (Figure 13-34).

This truss is made using the same two straw lengths. In this case, the tetrahedrons use the shorter length of straws, rather than the longer one. The longer straws are the ones that run vertically in Figure 13-34. Try some variations on this theme and see what kinds of strong structures you can make!

People also do exercises where they make trusses out of mini-marshmallow joints and dry spaghetti, or of course, there are lots of construction toys that include, or allow you to make, trusses. We encourage you to explore further!

THE GIFT SHOP

Our museum tour (and this book) has to come to a close at some point, and there are always more things to look at than there is time to cover. In this final section, we will give you a few more models, and also some ideas and jumping-off points that you might want to use as the basis for your own projects. Let's wander around and see if we can pick up a few things to play with at home.

THE TETRAHEDRON AND CUBE PUZZLE

Let's start with a 3D printed pair of models that will give you another way to think about the geometry of the cube and tetrahedron we just created. We can create a tetrahedron and an appropriately-sized hollow box, open on one side, and fit the tetrahedron diagonally into the cube.

We created the two files **puzzlebox.scad** and **puzzletetrahedron. scad** to make a cube and tetrahedron that will fit together, the way the models of straws did. The edge length of the tetrahedron is slightly less than times the inner dimension of the edge of the cube. The two models allow for clearances and all that automatically as long as you keep the variable **size** the same in both.

It looks like the tetrahedron can't possibly fit in the cube, since its

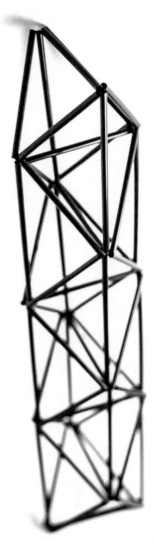

FIGURE 13-34: A bigger truss project

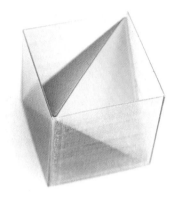

FIGURE 13-35: Inserting the tetrahedron in the cube

FIGURE 13-36: The tetrahedron in the cube

edges are about 40% longer than those of the cube. But if you line up a tetrahedron edge with the diagonal of the opening in the cube (Figure 13-35) it will slide in easily (Figure 13-36). This is a really fun puzzle to just hand someone, and it is a way to really appreciate the Pythagorean Theorem after you've spent 5 minutes fuming at not being able to get the tetrahedron in the cube.

REULEAUX AND CONSTANT-WIDTH SHAPES IN 3D

Now, let's check out some other interesting 3D shapes. Back in Chapter 4, we learned about Reuleaux polygons, which are shapes of constant width. What happens in the third dimension? Are there 3D shapes of constant width? It turns out there are several. A technical term for constant-width 3D shapes is a *spheroform*. We'll just call them constant-width shapes here, since we think that is more descriptive.

REVOLVED REULEAUX POLYGONS

First, think of the Reuleaux polygons we met in Chapter 4. Any of them could be folded along a line passing through the center and one vertex, and the sides would line up on top of each other. That line is called an *axis of symmetry*, a line that you draw such that a shape is identical (but reflected, like in a mirror) on either side of it.

Now imagine you take any of the Reuleaux polygons and spin them around that axis. A mathematician would say *revolve* it around the axis, and, if we revolve it so we get a closed 3D figure, the figure is called a *surface of revolution*.

It turns out that if we revolve any Reuleaux polygon 90 degrees (so that it fills all the space around its axis) it makes a 3D shape of constant width. Any cross-section will just be the original polygon, for which this is obviously true. If it is true for a cross-section, and if the cross-section is the same

FIGURE 13-37: Left to right: revolved Reuleaux triangle, pentagon, heptagon, nonagon

everywhere (because you spun the Reuleaux polygon around its axis of symmetry) then it is true everywhere. The model **revolved_reuleaux.scad** creates these models. It has the variables:

- `width = 50; //constant width`
- `reuleaux = 3; // number of sides, odd numbers 3 thru 13`

Figure 13-37 shows the figures generated **revolved_reuleaux.scad** by rotating the three-sided (triangle), five-sided (pentagon), seven-sided (heptagon) and nine-sided (nonagon) Reuleaux polygons.

When you roll any of these shapes on a flat surface, the highest point of each will remain at the same height, just like a sphere. Unlike a sphere though, the center of mass won't be equidistant from all of the surfaces, so the shapes will settle so that their center of mass is at the lowest possible point, rather than continuing to roll. This is a convenient feature if you need rollers that you would like to roll freely, but not go careening off every which way.

REULEAUX TETRAHEDRON

Another shape that is *very close* to a constant width is a Reuleaux tetrahedron. A Reuleaux tetrahedron is constructed similarly to a Reuleaux triangle, but in 3 dimensions. You make one out of four equal spheres, each with its center placed where the surfaces of the other three spheres meet. The area where all four spheres overlap looks like a tetrahedron, but with its edges and faces bowed outward, like the sides of a Reuleaux polygon. Here is an OpenSCAD model for a basic Reuleaux tetrahedron made up out of four 50mm spheres.

FIGURE 13-38: Reuleaux tetrahedron (blue) and Meissner type 1 (silver) and 2 (red)

```
size = 50;//radius of the spheres

$fs = .2;
$fa = 2;

intersection_for(i = [0:3]) translate(size / sqrt(2) / 2 * [(i
% 2) ? 1 : -1, (i == 1 || i == 2) ? -1 : 1, (i > 1) ? 1 : -1])
sphere(size);
```

Unlike the revolved Reuleaux polygons, a Reuleaux tetrahedron isn't a true shape of constant width. It is very close, but when it stands up on one of its edges, the opposite edge gets slightly higher than it should be if the width was constant Thus, the Reuleaux tetrahedron's largest dimension will be a little over **size** mm.

MEISSNER TETRAHEDRONS

A Meissner tetrahedron (named after mathematician Ernst Meissner, who published it in 1911) is a modified Reuleaux tetrahedron that has three of its edges rounded-off to make the width constant. Each rounded edge needs to be opposite an unrounded one. These can be either the three edges that share a vertex (which will be generated with a value of **meissner = 1** in **meissner.scad**), or three edges of the same face (**meissner = 2**). If you set **meissner = 0**, it will create a Reuleaux tetrahedron using spheres of **radius = size**. You can see all three in Figure 13-38.

```
size = 50; // radius of the spheres, and also the constant
width
meissner = 1; // Meissner tetrahedron (1 or 2)
               // or 0 for Reuleaux tetrahedron
```

PRINTING THESE MODELS

The various constant-width objects we have been talking about are a little

FIGURE 13-39: Support on a constant-width solid

FIGURE 13-40: Meissner solid rotates freely in a cube

tricky to print. Since all of the surfaces are curved, they require some amount of support when printing. We have been printing ours hollow, and usually have one of the edges pointed downward to minimize support. When they are hollow they will be lighter, so you are more likely to get away with printing them on an edge. Figure 13-39 shows how one of our prints came off the printer.

Another strategy for printing these shapes would be to cut them in half. This would give you a large, flat surface that each half can be printed on. After printing, you would just need to glue the two halves together.

BOUNDING BOX

In Chapter 4 we saw that Reuleaux polyhedra would rotate freely in a bounding polygon of one more side. We didn't point out that all the polyhedra will rotate freely in a square (although the higher-number-of-sides ones will not go as far into the corner of the square). You can try this by extension in a cube. If you use **puzzlebox.scad** to create a cube of the same internal dimension as the constant width of the solids, they will rotate freely (Figure 13-40). (Be sure to leave enough clearance.)

DUAL POLYHEDRONS

Dual polyhedrons are defined as polyhedra that will fit, one inside the other, such that the vertices of one are in the exact center of the faces of the other. This works for just a few polyhedra.

It turns out that an octagon fits inside a cube whose edge is **1/sqrt(2)** of the length of an edge of the cube. (When we 3D print, we also have to allow a little extra room, of course, so you would make the octagon a millimeter or

FIGURE 13-41: The cube-octahedron dual

two smaller.) Figure 13-40 shows this dual. (You can prove to yourself that this is right by drawing the triangle formed by the line from one vertex of the octagon to the other.) **1/sqrt(2)** is roughly 0.707.

You can make this using the open, hollow cube **puzzlebox.scad** model with its default internal side length of 50mm and the **edge_platonic_solids.scad** model (introduced in Chapter 3) with

```
edge = 50 / sqrt(2);
```

and, as described in Chapter 3, all the functions except the one making an octahedron commented out.

Placing the octahedron is tricky. It will want to slide around. We put a little bit of putty on the bottom center of the box to hold that vertex in place.

The octahedron and cube are duals of each other. That means that you can also make a cube with vertices that touch the center of every face of an octahedron surrounding it. That, however, would be tricky to 3D print. If you wanted a significant project to attempt, you could try printing an octahedron in two halves, and then inserting the (appropriately-sized) cube. Icosahedrons and dodecahedrons are also dual to each other, but that would also require that you create the outer polyhedron in two parts so you can insert the internal part of the pair.

A tetrahedron is self-dual, which means you can make a tetrahedron that fits inside another. You could use the pyramid volume model **pyramid. scad** from Chapter 8 to create two hollow tetrahedrons to try it! Remember to allow for the thickness of the outer tetrahedron; OpenSCAD dimensions are for the outside of the shape. We will leave you to play with how to explore these other combinations!

FIGURE 13-42: Golden ratio model

THE GOLDEN RATIO, PHI

If you look up the area of a pentagon in some books, you may see an odd-looking symbol, ϕ, in the formula for its area. It is the Greek letter *phi,* pronounced either "fie" or "fee" depending on the part of the world where you learned to say it.

The diagonal of a regular pentagon (a line connecting two non-adjacent vertices) is phi times an edge of the pentagon. Five-fold symmetry relationships, like the area of a pentagon, are often given in terms of phi.

Phi, like pi, is an irrational number (approximately 1.618). Phi can also be written in terms of the square root of 5:

$$\phi = (1 + \sqrt{5})/2$$

Phi is also sometimes called the *golden ratio,* since proportions of 1 to phi are particularly pleasing and come up often in nature. Let's say that I have any two numbers, a and b, that are in the golden ratio to each other. Then it turns out that:

$$a / b = (a + b) / a = \phi$$

The model **golden_ratio.scad** shows a series of boxes whose sides are in the ratio ϕ:1. Each smaller box has a longer side equal to the smaller side of the previous box. In other words, each box is scaled down by phi from the previous one, and rotated 90° (Figure 13-42).

FIGURE 13-43: A flower with 5-fold symmetry

THE GOLDEN ANGLE

Let's say we have a circle of circumference $1 + \phi$. If we wanted to break it up into two arcs that are in the golden ratio to each other, the smaller arc would have the internal angle of (approximately, since ϕ is irrational):

$$\text{Angle} = 360° \; (1/(1 + \phi)) = 360° * 0.382 = 137.5°$$

This angle is called the *golden angle*, and it arises, too, in many situations. Because the basis of this separation is an irrational number, if you start rotating around a circle by multiples of the golden angle, you will never quite get back to the place you started. Plants often (but not always!) have their subsequent petals on a flower arranged at the golden angle, so that they get more sunlight and don't block each other (Figure 13-43).

If you want to learn more, we have a chapter making 3D models of idealized plants and flowers using the golden angle in our book, *3D Printed Science Projects* (from Apress, 2016).

RELATION TO FIBONACCI SERIES

There are many projects you can do about phi. You can look up the *Fibonacci series*, which is a special list of numbers where each number is the sum of the last two. In other words,

Fibonacci series = 1, 1, 2, 3, 5, 8, 13, 21, ...

As the numbers get bigger, the ratio between subsequent ones gets closer and closer to phi. For example, 13/8 = 1.625, 21/13 = 1.6154, and so on. You might research the many situations in nature when the Fibonacci series arises, and see if you can find ratios equal to phi there too!

MORE THAN 3D: POLYTOPES

Throughout this book so far, we have talked about three (or fewer) dimensions. Now that we are at the very end, we want to be sure that you know that there are higher dimensions, too. Once you go beyond 3D, shapes are called *polytopes*.

In Chapter 3, we learned about Euler's characteristic, which connects the number of edges, faces, and vertices in a 3D object. Ludwig Schläfli (a Swiss mathematician working in the mid-to-late 1800s) first started thinking about how Euler's characteristics could be extended above 3 dimensions. We can fold 2D shapes to make 3D ones (as we saw in Chapter 9 when we learned about nets) and it turns out that you can do the equivalent, folding up 3D shapes to make 4D ones.

That's a little beyond what we want to do in this book, but we wanted you to know it is possible. There's a great video tying together the Platonic solids in 3D with (challenging!) 3D printed projections of 4D regular polytopes. Watch the episode "Perfect Shapes in Higher Dimensions" of the Numberphile YouTube channel, with computer scientist Carlo Séquin, by doing an internet search on "Carlo Sequin Numberphile" for that and more fun videos with Professor Séquin. We hope this opens up whole new dimensions for you to explore!

SUMMARY AND LEARNING MORE

If you want to learn more about Gothic architecture, check out the weighty (and wonderfully illustrated) book, *Gothic Architecture* by Paul Frankel; we used the 2000 edition revised by Paul Crossley, from Yale University Press. In the wake of the 2019 fire that severely damaged Notre Dame de Paris cathedral, there were also many popular discussions of how such buildings

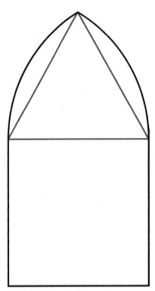

FIGURE 13-44: The geometry of the Gothic arch

were constructed. You might find more materials with some internet surfing. The New York Times had an illustrated tour on July 18, 2019, by Elian Peltier et al, entitled "Notre Dame came far closer to collapse than people knew. This is how it was saved." (Search on the author and title to see if it is still available online when you read this.)

To learn more about trusses in detail, you really need a bit of calculus and some physics. However, you can try looking up "trusses" and "free-body diagram" if you want to see the basics.

Finally, we hope you will explore some of the concepts we put forward in the last section. Any of them could be explored in many directions, from art to botany. We hope you will explore them (and everything else in the book) in as much detail as you have time for.

We appreciate you taking this time with us. We have an Appendix that follows this chapter with a few more resources, and a summary of how one might tie book concepts to the United States Common Core standards.

ANSWERS
GOTHIC VS. CIRCULAR ARCH

Let's call the width of the two arches at their *base* the variable base. The height of a circular arch of radius base / 2 is, by definition, just base / 2. Its perimeter is just half the circumference of the circle, or π base / 2.

What about a Gothic arch, though? If we remember that we have an equilateral triangle inscribed in the arch (see Figure 13-44) then we know that all the angles of that triangle are 60°.

That means: sin(60°) = height / base

We can look up sin(60°) and discover it is 0.8660, which also happens to be $\sqrt{3}$/2.

The height of a Gothic arch then is $\sqrt{3}$/2 * base (versus ½ * base for a circular arch), so the Gothic arch is $\sqrt{3}$ times taller than a circular one of the same base width.

What about the perimeter? We know we created it with two 60° arcs, each with a radius equal to base. 60° is 1/6 of a circle.

The length of a 60° arc of a circle of radius *base* = 2 π *base* / 6 = π *base* / 3
There are two of these arcs in our arch, so the perimeter of our arch is:

Perimeter = ⅔ π *base* (versus ½ π *base* for a circular arch).

Thus a Gothic arch will take about 33% more stone than a circular one, for a 73% taller arch.

APPENDIX: RESOURCES AND STANDARDS

In this Appendix, we summarize key resources we think of as companions to this book, as well as other sources we found particularly useful. Finally, we will summarize how we imagine teaching from this book, and where we feel the book intersects with the Common Core math instructional standards.

COMPANION RESOURCES

The models described in this book are available under a CC-BY 4.0 Creative Commons license, at this Github repository: **https://github.com/whosawhatsis/Geometry**

See Chapter 2 for details on downloading them. This repository contains all the models used in the book, as well as a few models not described in this book. Some of the models support lesson plans for visually impaired students, or other projects. You are welcome to explore these other models as bonus adventures.

We have also developed companion lesson plans that use some of the same models as this book. There is also a Teacher's Guide to get you started. These are designed for teachers of visually impaired students, but other educators are welcome to explore them as well. Some of the models are unique to these lesson plans to support these learners, but perhaps will be of interest to others as well. (Notably, we have significantly different conic section models.) You can find the lesson plans, Teacher's Guide, and more at **https://www.nonscriptum.com.geometry**.

We used the OpenSCAD computer-aided design program to create our models. This free software, and an excellent manual, are available for Mac, Windows, and Linux (but not tablets) at: **https://www.openscad.org**. OpenSCAD recently had a very significant upgrade to version 2021.01. The models were created with a mix of 2021.01 and the predecessor version (2019.05) and we have endeavored to keep back-compatible with 2019.05 where possible in case you are not able to upgrade to the latest version.

OTHER SOURCES

There are many excellent web sources for mathematics. For the most part, we have avoided derivations and extensive algebra. If you would like to be talked through the concepts of this book in a more traditional way, we highly recommend the Khan Academy (**www.khanacademy.org**).

Wikipedia (**www.wikipedia.org**) is also pretty solid as a resource for most mathematics at this level, and we frequently cite particular articles as we go in this book. The Math is Fun website (**www.mathisfun.com**) is solid as well.

There are also many YouTube mathematicians (although not always aimed at the K-12 level). At the more esoteric end, we particularly love the work of 3Blue1Brown, who has a YouTube channel of that name.

We also appreciate Paul Lockhart's books on teaching math in a hands-on way, particularly his *A Mathematician's Lament* (Bellevue Literary Press,2009) and *Measurement* (Belknap/Harvard University Press, 2012). Jo Boaler's **Mathematical Mindsets** (Jossey-Bass, 2016) is in a similar vein.

Finally, the author of OpenSCAD has just co-authored a book on learning to code with OpenSCAD, *Programming with OpenSCAD: A Beginner's Guide to Coding 3D-Printable Objects* by Justin Gohde and Marius Kintel (2021, No Starch Press). You might find that a good resource for more depth on OpenSCAD's ins and outs than we provide.

TEACHING FROM THIS BOOK

The following table suggests which Common Core standards might be covered by the material for some of the chapters. The geometry standards for grades K-12 can be found at **http://www.corestandards.org/Math/Content/G/**. Since we have selected topics we felt particularly benefited from 3D models, it is not a complete, end-to-end curriculum for middle or high school geometry. Also, since many schools now spread out geometry across various grade levels, any alignment by grade level is by nature going to vary by location. Different topics can be taught more simply or experimentally at younger ages, and more algebra than we do here can always be added in or tied to the models.

There is always some interpretation in thinking about which projects illustrate which standards, and we recommend teachers review the material and adapt it as needed to their interpretation of the standards they need to cover.

In many cases, we have gone for physical models over equations to build intuition, so it is open to interpretation if the material meets the standard that says to derive the equation via reasoning. For example, we show students how to use a directrix and a focus to construct a parabola, but we stop short of deriving the equation. We feel these derivations are covered in depth in dozens of other existing sources and would be a distraction here.

We also list keywords for topics we cover in each lesson, which we thought might be more useful for homeschoolers. The Table of Contents provides a more thorough listing of topics.

Some topics, like learning to do computer programming, do not naturally fit into the math standards. We also include some physics, astronomy, and architectural applications. Some of these are somewhat open-ended, so we just note those chapters as broad applications of the concepts in the book.

Finally, we have material in the book that is not part of the core math curriculum but which we feel ties together some of the math concepts and makes them concrete, like determining latitude and longitude, or learning about buoyancy forces. We don't attempt to shoehorn these topics into any existing Common Core topic, but hope you will explore these hands-on applications with your students.

CHAPTER	CHAPTER TITLE	TOPICS	GRADE LEVEL	RELATED COMMON CORE STANDARDS
1	What is Geometry?	Define and introduce Euclidean geometry	varies	
2	OpenSCAD and Math Modeling	Learning to code	varies	
3	Simulating Geometry	Circles, pi, angles, polygon, polyhedron, platonic solid, Euler characteristic (sometimes Euler's Formula), transformations. Also more on learning to code.	1,2,8	CCSS.Math.Content.1.G.A.1, CCSS.Math.Content.2.G.A.1 CCSS.Math.Content.8.G.A.4,
4	Constructions	Construct angle bisector, perpendicular bisector, equilateral triangle, Reuleaux polygons	HS	CCSS.Math.Content.HSG. CO.D.12
5	The Triangle Bestiary	area of a triangle; scalene, isosceles, equilateral, right, acute, obtuse; angles of a triangle; congruent and similar triangles	1, 4,6, 8	CCSS.Math.Content.1.G.A.1, CCSS.Math.Content.1.G.A.2, CCSS.Math.Content.4.G.A.2, CCSS.Math.Content.6.G.A.1, CCSS.Math.Content.8.G.A.4, CCSS.Math.Content.8.G.A.5,
6	Pythagoras and a Little Trigonometry	Pythagorean theorem, sine, cosine, tangent, angles above 90 degrees	8, HS	CCSS.Math.Content.8.G.B.6, CCSS.Math.Content.8.G.B.7, CCSS.Math.Content.HSG. SRT.C.6, CCSS.Math.Content.HSG. SRT.C.7 ,CCSS.Math. Content.HSF.TF.A.1
7	Circles	Area of a circle; inscribed and circumscribed polygons; estimating pi; latitude and longitude	7, 8	CCSS.Math.Content.7.G.B.4 CCSS.Math.Content.8.G.B.8 Real-world applications of concepts

8	Volume	Volume of rectangular solid, prism, cylinder, cone, sphere, pyramid, Cavalieri's principle, density, displacement, Archimedes' principle, buoyancy	6,HS	CCSS.Math.Content.6.G.A.1 CCSS.Math.Content.6.G.A.2 CCSS.Math.Content.6.G.A.3 Real-world applications of concepts, including CCSS.Math.Content.HSG.MG.A.2
9	Surface area and Nets	Nets, surface area of platonic solids, right prisms, pyramid, cone, cylinder, sphere	6, 7, HS	CCSS.Math.Content.6.G.A.4, CCSS.Math.Content.7.G.B.6 CCSS.Math.Content.HSG.GMD.A.1
10	Slicing and Sections	Slicing cubes, prisms, cylinder; understanding resulting cross-sections; conic sections	7, HS	CCSS.Math.Content.7.G.A.3, CCSS.Math.Content.HSG.GMD.B.4
11	Constructing Conics	Conic sections, directrix and generatrix constructions, Kepler's laws, real-world applications of conic sections	HS	Standards are defined in terms of using these techniques to generate an equation, but we use them to draw curves instead. Related applicable standards are CCSS.Math.Content.HSG.GPE.A.2 and CCSS.Math.Content.HSG.GPE.A.3
12	Geometry, Space and Time	Equation of Time, orbits of planets, applications of conic sections, and trigonometry	varies	Applies techniques developed throughout the book.
13	The Geometry Museum	Applying geometry principles to real-world problems	varies	Applies techniques used throughout the book.

INDEX